ECOLOGICAL GOVERNANCE

ECOLOGICAL GOVERNANCE

Toward a New
Social Contract with the Earth

BRUCE JENNINGS

WEST VIRGINIA UNIVERSITY PRESS

MORGANTOWN · 2016

First edition published 2016 by West Virginia University Press
Printed in the United States of America

24 23 22 21 20 19 18 17 16   1 2 3 4 5 6 7 8 9

ISBN:

cl   978-1-943665-15-0
pb   978-1-943665-18-1
epub   978-1-943665-17-4
pdf   978-1-943665-16-7

Library of Congress Cataloging-in-Publication Data is available
from the Library of Congress

Book and cover design by Than Saffel

To Strachan Donnelley

(1942–2008)

philosopher, outdoorsman,

conservationist, philanthropist,

teacher, friend

The sum total of harm inflicted on the world so far equals the ravages a world war would have left behind. . . . We so-called developed nations are no longer fighting among ourselves; together we are all turning against the world. We shall thus seek to conclude a peace treaty. . . . That means that we must add to the exclusively social contract a natural contract of symbiosis and reciprocity in which our relationship to things would set aside mastery and possession in favor of admiring attention, reciprocity, contemplation, and respect; where knowledge would no longer imply property, nor action mastery, nor would property and mastery imply their excremental results and origins.

—Michel Serres[1]

Contents

Part IV: The Political Economy of Climate Change— Democracy, If We Can Keep It

Introduction

The global economy today is overwhelming the ability of the earth to maintain life's abundance. We are getting something terribly wrong. At this critical time in history, we need to reorient ourselves in how we relate to each other and to the earth's wonders through the economy. We need a new mass movement that bears witness to a right way of living on our finite, life-giving planet.

—Peter G. Brown and Geoffrey Garver[1]

Solving our problems in the time we have available is not possible if all we do is change our technology. We will not arrest ecological decline or regain financial health without also introducing a different rhythm of work, consumption, and daily life, as well as alternation in a number of system-wide structures. We need an alternative economy, not just an alternative energy system.

—Juliet Schor[2]

Our entire economic system is fundamentally dependent on the functional integrity of natural and living systems that are losing patience with us. That is to say, these systems have a limited capacity to tolerate human extraction from them and excretion of waste products and by-products into them. Today human economic activities worldwide are approaching those limits; in some cases they may have already exceeded them. We won't necessarily realize that we have exceeded them right away. The consequences may be delayed, subtle, and systemic rather than clearly localized and visible. In short, our situation is insidious. Our excess does not lend itself to the age-old human ways of perceiving danger and taking appropriate, timely steps to protect against it. Our excess is the most grave danger we face.

In what he called a "land ethic," the noted American conservationist Aldo Leopold said that a thing is right when it tends to preserve the integrity, stability, and beauty of the biotic community. The land ethic is a call for human beings to be at peace with the planet, to live in a place without spoiling it, to accommodate our technology and ingenuity to the needs of other forms of life as well as to our own, to respect natural limits, and to keep ourselves within safe operating margins that do not violate planetary boundaries and tolerances.

Today organized economic activity all over the world is assaulting natural integrity, stability, and beauty, not always with results that are immediately apparent and not always intentionally, but pervasively, persistently, and with devastating cumulative effects. Humanity is at war with the planet. The time has come when all human minds, all over the world, must focus on a peace treaty to end this unwinnable war. In her 2014 study of climate change, Naomi Klein perceives that endless economic growth and consumption is a war we cannot win: "our economic system and our planetary system are now at war. Or, more accurately, our economy is at war with many forms of life on earth, including human life."[3] And many of the world's leading scientists, who issued a sobering 2013 report on the state of knowledge and research on climate change, remind us that those now waging the war will not be its ultimate casualties:

> [A] set of actions exists with a good chance of averting "dangerous" climate change, if the actions begin now. However, we also know that time is running out. Unless

a human "tipping point" is reached soon, with implementation of effective policy actions, large irreversible climate changes will become unavoidable. Our parent's generation did not know that their energy use would harm future generations and other life on the planet. If we do not change our course, we can only pretend that we did not know.[4]

I am writing this book with the following question in mind: If fifty years from now our children's children could ask us why we did what we are doing, what could we tell them? Pretending that we did not know is a shameful legacy.[5] Arguably, the actions of every generation pass on problems and burdens to the next, but the shadow on the future we are creating in our time may be historically unprecedented in its severity and in the disruption to all life on earth it will bring about. Yet, there is still time to redeem our generation and to lighten the burden that we shall bequeath to the next. In order to do so, the economic activities of humankind, most especially those in the most affluent societies, will require new aims, values, and modes of governance.[6] Achieving this future will require what I call an ecological social contract.

Nature as Stock and Sinkhole

Why is our collective activity, especially economic activity, colliding with ecological system limits and encroaching upon the safe operating margins of the planet? The reasons are many, but one key factor is that we think of the human realm as set apart from the rest of the world, and we believe that we can manipulate nature, engineering it as we see fit in accordance with what we find meaningful and valuable. We seem blindly determined to pursue this viewpoint to its logical extremes. Biophysical systems, even when they are scientifically well understood, are mistakenly seen as *things we live off of*, not as *places we live within*. For the most part human economic activity is conducted as if nature were just a stock of raw materials and energy humans consume, and as if nature were an endless dumping ground (a "sink") into which we put our waste products. We extract useful, energy-rich materials from nature and excrete useless, degraded by-products into it. We take in and we throw away. Some of what we take in is running out, such as the once-teeming ocean fisheries, and the places where we throw things away are filling up and becoming overloaded, like leaching landfills or rising levels of carbon dioxide in the atmosphere. In reality there is no such place as "away"; there is only a shifting of cost and burden to another place and to someone or something else.

The idea that the planet is a stock and a sinkhole is so widespread that it forms an unnoticed background assumption, not only of mainstream economic thought, but also of many habits and customs in our daily lives. In his 2015 encyclical, *Laudato Si': On Care for Our Common Home*, Pope Francis reminds us of how this was not always the case and of how dire a change in human orientation this kind of thinking represents. As Bill McKibben notes, the pope's teaching here is much more than a contribution to the climate debate. It is a "sweeping, radical, and highly persuasive critique of how we inhabit this planet—an ecological critique, yes, but also a moral, social, economic and spiritual commentary."[7]

Specifically, Francis ties together the question of human freedom with the question of setting limits and repositioning ourselves and our activities through a restructuring of our basic convictions and contentments. "Each age tends to have only a meagre awareness of its own limitations," he writes. "It is possible that we do not grasp the gravity of the challenges now before us. . . . Our freedom fades when it is handed over to the blind forces of the unconscious, of immediate needs, of self-interest, and of violence. . . . [W]e cannot claim to have a sound ethics, a culture and spirituality genuinely capable of setting limits and teaching clear-minded self-restraint." He then focuses on technology as a power that mediates the relationship between humans and nature and shapes how we inhabit the planet:

The basic problem goes even deeper: it is the way that humanity has taken up technology and its development according to an undifferentiated and one-dimensional paradigm. This paradigm exalts the concept of a subject, who, using logical and rational procedures, progressively approaches and gains control over an external object. This subject makes every effort to establish the scientific and experimental method, which in itself is already a technique of possession, mastery and transformation. It is as if the subject were to find itself in the presence of something formless, completely open to manipulation. Men and women have constantly intervened in nature, but for a long time this meant being in tune with and respecting the possibilities offered by the things themselves. It was a matter of receiving what nature itself allowed, as if from its own hand. Now, by contrast, we are the ones to lay our hands on things, attempting to extract everything possible from them while frequently ignoring or forgetting the reality in front of us. Human beings and material objects no longer extend a friendly hand to one another; the relationship has become confrontational. This has made it easy to accept the idea of infinite or unlimited growth, which proves so attractive to economists, financiers and experts in technology. It is based on the lie that there is an infinite supply of the earth's goods, and this leads to the planet being squeezed dry beyond every limit.[8]

Despite the fact that this outlook is based on a lie, as the pope maintains, or on a serious form of misprision, our institutional system is set up to function well when we follow these habits unreflectively and when our behavior is deliberately strategic so that we stay within the rules of the system to pursue our own competitive advantage and individual or group self-interest. We go with the flow of economic activity, and we game the legal, market, and political systems to our own advantage (if we can), but always in ways that nonetheless sustain that flow. In this sense we are parties to a vast and tacit agreement, a social contract of consumption.

Of course, no one can think explicitly about all background assumptions or implicit perspectives all of the time. There are historical moments, however, when the tacit needs to become explicit, the pre-reflective should be revealed and reflected upon. We are in such a moment right now. The social contract of consumption can be reconstructed as a new ecological social contract. The current psychological and economic defaults of individualistic strategic thinking—namely, "What's in it for me?"—must be reset to relational ethical thinking that is mindful of human interdependence, sustaining the natural commons, and promoting the social common good, such as "What's in it for diverse, abundant, and resilient life?"

This book is a reflection on this reconstruction of the social contract and its prospects for success. I do not have a checklist of specific new policies and practices to offer. I am not the person to do that, and besides, there is a wealth of such technical knowledge and creative problem-solving ideas now available.[9] What is also needed, but harder to find, are discussions of those fundamental ideas and concepts offering ethical justification for acting—and the democratic political will to act—on what science knows and what humanity, properly understood, requires. My contribution is to that discussion.

In the Western political tradition the ethical justification for doing what humanity requires has often been explored and shaped by using the metaphor of the "social contract." It helps us think about the following fact: We need others in order to survive, but we aspire to be free from the constraints and coercion others impose on us. The resolution of this conundrum is the autonomous agreement by each person to freedom-limiting common rules that fairly benefit all and unfairly burden none. Mutual agreement and promise-making (contracting) is the key to resolving the paradox that in order for the individual freedom of each to be sustained, it must be justly limited by all.

Reconciling order and freedom is an enduring challenge, never more so than today. The social contract idea can be a lens through which we better perceive our own situation and options. By exploring the terms of a relational, ecological agreement, we can reconcile desires that are unlimited with true powers and capabilities that are limited. Accordingly I have chosen the metaphor of the social contract as a point of entry. The destination is an ecological and relational understanding of "political economy," that is, the intersection of production, distribution, and governance, which represents the most consequential way that human beings relate to natural systems.

The pathway between this starting point and destination is straightforward. I begin in Part I with an interpretation of the idea of a social contract, the philosophical assumptions about human beings it rests on, and the normative functions it performs. I turn next to the idea of a political economy, which I regard as an institutional and practical manifestation of the social contract: a consumptive social contract will give impetus to one form of political economy, an ecological contract will foster another.

With the basics of the distinction between the consumptive and the ecological contracts in place, in Part II I take up the theme of what I call "nature in humans." I do this with a series of reflections on the political philosophies of Thomas Hobbes and Jean-Jacques Rousseau, the two thinkers who, in my estimation, most clearly show the political and moral potential inherent in the idea of a social contract. They did not use this idea merely to sketch out a minimalist modus vivendi. They got behind the idea of individuals' free submission to social order and common rules to explore a fundamental transformation in humans—from "natural" or wild freedom to "artificial" or domesticated obedience and orderly conduct. This is the achievement of a second nature in human beings, a political and moral nature, with motivations and capabilities for rational thinking that supersede their first or pre-political and anarchical way of being. In pursuing this, Hobbes and Rousseau importantly diverged: Hobbes laid important foundations for what later became the consumptive social contract, with his notions of competitive individualism driven by unlimited desire. For his part, Rousseau glimpsed the shape of a new ecological social contract with his insights into the dynamic and co-evolving connection between nature, culture (symbolic orders of meaning), and the human mind.

From the theme of nature in humans I turn in Part III to the activities of humans in (and on) nature. I begin with a discussion of the aspects of human action and agency that are key to making a transition to a new social order based on an ecological vision and conscience. With the differences between the consumptive social contract and the ecological social contract in place, I next explore more deeply what I take to be the key provisions or terms of the ecological social contract. These terms are: (1) creating a system of wealth that moves beyond affluence and scarcity to plenitude and frugality; (2) creating a system of property that moves beyond commodities to commons; (3) creating a system of freedom or liberty that moves beyond noninterference and independent self-sufficiency to relationality and interdependence; and (4) creating a system of citizenship that moves beyond self-interested advantage to trusteeship for right relationship and right recognition.

The sphere of right, or justice, as I understand it, includes (1) the moral community of both human and nonhuman life and (2) the ecological commons—the symbiotic, interdependent systems, both natural and social, upon which human being and becoming depend.

In Part IV I close with a discussion of how an ecological social contract can illuminate the global problem of climate change. This problem sets in relief the political and governance aspects of the idea of a social contract. A particular form of democracy—namely, interest group representative democracy—has heretofore been the political arm and twin brother of the social contract of consumption. Can effective and timely climate change governance succeed in a democracy fundamentally dependent on consumption and continuing economic growth? I think not. That leaves the alternatives of abandoning democratic governance in favor of some kind of ecological authoritarianism, on the one hand, or reconstructing democracy through the exercise of ecological citizenship and trusteeship, on the other. If we move into an ecological authoritarianism, we should do so with open eyes rather than with frightened hearts. For my part, I aspire to and defend the governance of an ecological democracy. But at the very least, I hope that this book will help us to think carefully about the choices and challenges we face.

Part I

Rethinking Life, Liberty, and the Pursuit of Happiness on a Planet in Crisis

The era of procrastination, of half-measures, of soothing and baffling expedients, of delays, is coming to its close. In its place we are entering a period of consequences.

—Winston Churchill[1]

Certainly there is a necessary time and place in human existence for the so-called "ethical phenomenon," that is to say, the experience of obligation, the conscious and deliberate decision between something which is, on principle, good and something which is, on principle, evil, the ordering of life in accordance with a supreme standard, moral conflict and moral resolve.

—Dietrich Bonhoeffer[2]

1

The Social Contract

As it was when Churchill spoke on the eve of World War II at a time of Nazi conquest and genocide, we are again living in a period of consequences when procrastination and baffling expedients must come to a close. What are the imperatives of human responsibility in a time when virtually all things on the planet, both living and nonliving, are materially affected by human activity?

The question of human responsibility and flourishing is the fundamental and timeless concern of political and moral philosophy, but it takes practical shape in specific cultures and historical moments. The idea of a "social contract" is one way this question has been addressed beginning with the ancient Greek Sophists. In Plato's *Republic*, written in dialogue form, Socrates is speaking with a promising young man named Glaucon about the origins of political society and justice. Glaucon summarizes the viewpoint being taught by the Sophists:

They say that doing justice is naturally good, and suffering injustice is bad, but that the bad in suffering injustice far exceeds the good in doing it; so that, when they do injustice to one another and suffer it and taste of both, it seems profitable—to those who are not able to escape the one and choose the other—to set down a compact among themselves neither to do injustice nor to suffer it. And from there they began to set down their own laws and compacts, and to name what the law commands lawful and just. . . . The just is . . . cared for not because it is good but because it is honored due to a want of vigor in doing injustice. . . . So the argument goes.[1]

To give some indication of the persistence and influence of this notion of agreement, consider the following formulation by Antoine de Monchrétien in 1615, two millennia after Plato. Monchrétien was an important dramatist and author of an influential book on public policy, *Traicté de l'economie politique*. In this work one finds clear echoes of Glaucon's rendition of the Sophists' social contract. A self-centered and competitive view of human conduct is coupled with a historical and sociological way of thinking about human nature:

Since we are not perfect and do not live among perfect men, let us speak to this point [that ambition is "a powerful spur to doing well"] according to the way the world actually operates. It is a world in which each individual sets his sights on profit, and his eyes are caught by the glittering of each little spark of utility, to which a man is drawn, whether by nature or by the education and custom which we speak of as second nature. . . . Therefore the most skillful, who have studied most carefully the book of common experience, have held that the diverse necessities which each individual senses strongly as pertaining to himself have been the first cause of general communities. For the most ordinary between men, and their most frequent assembling together, depends on the help they provide for one another and the mutual offices they render to each other, in such fashion that each is primarily motivated by his

individual profit, as he perceives it, rather than by
nature, the real prime mover here, of which he is
unaware. So many efforts, so much labor of so many
men has no other goal but gain.[2]

Several important ideas are already present in these formulations, and they will come up again in this book as we go along. One is the notion that the political association created by the social contract is a mutual protection arrangement among those who are essentially self-interested, vulnerable to exploitation by others, but too weak individually to control and exploit others themselves. Closely related to this is the idea that justice is essentially a will to power, and the social contract is a collective will to power that protects against all unduly powerful individuals or groups. Self-interest under conditions of vulnerability and uncertainty provides a logic for obedience to political authority and self-restraint, but the hearts of the members of the social contract are not really in it. The social order provides them with contentment, but not conviction, and even that contentment is alloyed by the secret desire to dominate and control. Glaucon recognizes that this is an inadequate foundation for social order and challenges it by imagining a situation in which an individual (who possesses a magic ring of invisibility) could evade the collective logic and commit injustice—steal from, use, or otherwise take unfair advantage of others—with impunity. Free from the fear of punishment, why would anyone who was rational obey the common rules? Such a one may appear to enter into the social contract and abide by the common rules, but would not truly enter into that community of reciprocal promising and trust.

Perhaps even this brief glimpse of the social contract idea is enough to suggest how fecund it is for political philosophy. No wonder that over the centuries it has been revived again and again, particularly by thinkers living at times of atypical social change and ferment, when the fundamentals of authority, legitimacy, and the logic of collective action are raised explicitly. Europe in the seventeenth and eighteenth centuries went through such a time and saw a burst of creative social contract thinking. It has also been revived powerfully in several comprehensive political and ethical theories during the past fifty years.[3]

In this book I shall be using the idea of the social contract more informally as an orientation and a metaphor rather than as a conception in a more comprehensive theory. In common parlance the social contract refers to fundamental norms of common consent that provide social cohesion. It is the mutual agreement that provides grounds for our moral ideals and our sense of worth and happiness—in other words, our conviction and contentment. As such, the idea of a social contract is a useful way to think about our current situation and predicament.

Obviously, the social contract is not literally an agreement made by every member of a society at a particular time and place. Yet, one does recall a few historical examples that approximate that or at least got a process like that started: the *Magna Carta* in 1215; John Winthrop's sermon, "A Model of Christian Charity," on the *Arbella* bound for the shores of Massachusetts in 1630; the Continental Congress that issued the American Declaration of Independence in 1776; the Constitutional Convention that wrote a new Constitution for the United States in 1787; the Oath of the Tennis Court and the Declaration of the Rights of Man and Citizen in France in 1789; Abraham Lincoln's "Gettysburg Address" in 1863; and the passage of the United Nations Universal Declaration of Human Rights in 1948.

But in political philosophy the social contract is a hypothetical thought experiment, a "just so" story, a "what if" idea:

"What if we were starting a new society from scratch and constructing common rules for living together? What would those rules be?" In actual history, authority is set up gradually, rules are made, and people come to understand the rationale and justification of those rules and why they should be obeyed. But how can such historical facts be justified and what good reasons might support them? The best reason is that the rules are right and just and promote the good of human development and flourishing for those who live within them. Other reasons involve more pragmatic calculations of self-interest and mutual advantage. Still other reasons and motivations have to do with fear of life in the absence of such rules, or fear of consequences and punishment if one violates them and gets caught.

Another way to explain the idea of the social contract is to see it as a metaphor for the human condition, namely, the situation of fearful and venerable, yet intelligent and resourceful, relational beings, who are interdependent and need one another and who must survive cooperatively if they are to survive at all.

Still another way to view a social contract is to see it as a solution to the problem of how to combine freedom and social order, individual rights and common rules, for beings who cannot flourish without both. Near the beginning of his work on the social contract, Rousseau said:

> In the context of my subject, this difficulty can be stated in these terms: "Find a form of association that defends and protects the person and goods of each associate with all the common force, and by means of which each one, uniting with all, nevertheless obeys only himself and remains as free as before." This is the fundamental problem which is solved by the social contract.[4]

Taking these perspectives into account, a social contract can be defined as a pact for cooperative, mutually beneficial living together sustained by the performance of human actions of certain kinds and by the omission of others. These actions, in turn, are based on certain motivations that fall under the headings of motivations of conviction and motivations of contentment. We strive to do what is right and good and we desire to be safe, fulfilled, and happy with ourselves, our work, our relationships, and our lives.

One concrete expression of a social contract in this sense is the American Declaration of Independence, which articulated the rationale and normative goals of breaking away from England and becoming a sovereign nation-state in the 1770s. Written principally by Thomas Jefferson, it states the basis of a social contract in the following terms: "We hold these truths to be self-evident, that all men are created equal, that they are endowed by their Creator with certain inalienable Rights, and that among these are Life, Liberty, and the pursuit of Happiness."[5] Social contracts do not necessarily invent the basic ontological and moral rationales for a particular way of living; instead social contracts build on these foundations, inchoate and often threatened by circumstance, that are rooted in the condition of nature in humans and humans in nature.

Since those words were written, the American social contract that animates much of our economic and political projects and behavior has increasingly emphasized control and consumption as the basis for life, liberty, and happiness. Both in regard to material things (resources and goods) and to human capability (labor and services), each individual wants to control and use as much as possible in the pursuit of his or her desire and self-interest or advantage. I shall have much more to say about this interpretation of the social contract, this particular way of fashioning a social order and reconciling individual will and freedom with communal rules and authority. For now, let me simply name it as the *consumptive social contract or the social contract of consumption*.

This is a social contract in which ordinary citizens, the *demos*, have given elites political control in exchange for the promise and performance of economic growth and material affluence. The history and the practicality of this social contract correspond roughly to the era of the Industrial Revolution, and especially to the widespread use of technologies based on energy-rich fossil carbon during the last two centuries. The consumptive social contract has informed ethics, economics, politics, and governance during the period of significant achievements in human history—Enlightenment modernity, market and corporate capitalism, liberalism, and representative democracy. All these innovations have been built on the extraction of rich energy from fossil carbon. It has given the cultures of the North Atlantic, and increasingly those around the world, a normative vision of individual liberation, but we can now see that it has become a distorted and unsustainable framework. Exchanging democratic citizenship for private consumerism is no longer a good deal.[6]

Today the fundamental questions of human responsibility and flourishing are shifting and new shapes of practical responsibility are emerging. This is the time that many have begun to call the "Anthropocene," the Human Age. It requires a reconstructed modus vivendi or pact of living together—a new ecological social contract to replace the consumptive social contract. To take the notion of the Human Age seriously is not to disregard or detach from nature. It calls upon humankind to engage with nature more deeply, more intelligently, and more respectfully than ever before.

The ecological social contract is a new framework of interpretation—a new collective story of how life, liberty, and the pursuit of happiness should be normatively ordered and institutionally arranged. It can be prompted by making the familiar strange, thereby fostering the realization that the terms of the consumptive social contract are not inherent in the nature of things, unchangeable, or hardwired. These terms of how humans should be in nature are not dictated by nature in humans.

When it comes to responsibility, it is certainly true that not all human beings are equally responsible for, or equitably benefit from, the deleterious impacts of human activity on global ecosystems. Some, though, have been at war with the planet for a very long time, and today the damaging footprints of the global few are vastly larger than those of the global many. (This is a function of massive global inequality. Almost 50 percent of the world's wealth is held by 1 percent of the world's population. The eighty-five richest people have as much as the 3.5 billion poorest.[7])

Nonetheless, the terms and conditions of an ecological social contract challenge all humanity to think explicitly and critically about the taken for granted, to think anew about the responsible use of our power and technology in the face of growing scientific knowledge concerning ecological and biophysical realities and limits. Like the consumptive social contract that precedes it, but in fundamental and substantively different ways, an ecological social contract will inform the two principal well-springs of human behavior that I referred to earlier—cognitive *conviction* based on good reasons and rationales and affective *contentment* based on the effective striving for human flourishing and happiness. The coming ecological social contract will leave no one out.

The very notion of a social contract is based on the idea that human social and political order exists, as it were, naturally and necessarily because human beings are fundamentally interdependent creatures—"political animals (*zōon politikon*)," as Aristotle expressed it. We act for reasons, and we also have complex aspirations and psychological needs. Indeed, our needs and desires sometimes supply our reasons, while at other times our reasons nurture new desires and even needs. Both of these capabilities can easily lead to conflict and undermine social cooperation and peace. Hence, the proper ordering of our convictions and contentments is essential. Such order is made up of culturally meaningful rules, roles, and relationships that are almost universally understood and agreed to, explicitly or tacitly. The consumptive social contract offers one specific understanding of what the underlying rationale of social and political order is, and what its rules (formal and informal norms and patterns of behavior), roles (structures of authority, expertise, and activity), and relationships (modes of justice, freedom, and equality) should be. The ecological social contract does likewise, but with a conceptual and motivational content that will lead to different manifestations in human political, economic, and social behavior.

2

Political Economy

At the heart of both the consumptive and the ecological social contracts lies the domain of "political economy," which represents the mode of collective human activity with the greatest impact on natural and living systems. A political economy is a structure of power and a form of large-scale social coordination. It can function only if millions of people willingly fit into and participate in it without undue resistance, effectively reproducing it over time, with modifications but normally without fundamental transformation. Such behavior is the result of coercion, to some extent, but largely it grows out of a cultural perception that the political economy on offer is simply "the way things are" or "the way the world works." The rationale for any given political economy tends to be utilitarian and functionalist: a political economy is legitimate to the extent that it promotes social order, makes cooperative activity effective, creates wealth, and meets human needs and supports human happiness in the aggregate, even if the distribution of wealth and opportunity is highly unequal.

By continuing to ignore fundamental aspects of the natural world upon which human economic activity draws, the management and governance of contemporary political economies will gradually erode the capacities to perform these functions and to fulfill these goals, and they will lose their social and ethical legitimacy. That is to say, contemporary advanced political economies are not sustainable and are self-contradictory because they are undermining their own foundations—thermodynamic, natural, biological, social, and normative.

It is ironic that only two or three decades after the seeming global triumph of state capitalism over its main twentieth-century rival, state socialism, fundamental new challenges to it have emerged. One is the cultural and religious backlash against a secular worldview and its materialism evident in many parts of the developing world. However, the global political economy of state capitalism now faces an opponent far more formidable than either the international communist movement of old or the various religious and cultural revitalization movements today. The opponent now is natural reality itself—the reality of the second law of thermodynamics, the reality of the coming end of the fossil fuel era, the reality of atmospheric and oceanic perturbations resulting from greenhouse gas pollution generating global climate change. Aspirations to "defeat" this opponent, to defeat nature itself, sometimes take the form of a technological imagination (*Star Trek*), which can be alluring, and sometimes that of a technocratic fascism (*Avatar*), which is not. A more likely goal, but still far, far from certain, is détente with the reality of natural limits. But this future détente will not be like the Cold War détente with the USSR; it will not permit the flourishing of growth and capitalist development within it. The new peace treaty with the planet will call for a fundamentally new form of political economy.[1]

If our social contract of consumption and its political economy stand at a crossroad, which direction will they take? Will it be toward an accommodation to natural limits or toward yet another technological overcoming of them? After all, human structures of power have reshaped and redefined natural reality in the past, and overcome what were then taken to be natural limits. Perhaps the problem, then, is merely to channel sufficient resources (money, talent) into technological innovation. Behind this approach stands a narrative of human science and technology as a continuous process of *adapting* nature to human will and activity rather than *accommodating* human will and activity within the context of nature. Those who see accommodation as the only viable course embrace instead a historical narrative of transition and discontinuity. In their view, the technological and economic future will be qualitatively and not merely quantitatively different from the past; different not simply in degree, but in kind.

Have we reached at last a kind of thermodynamic and eco-systemic bedrock? I believe that we have. The narrative of adaptation has always been an exaggeration, an illusion born of the pride of power. Earlier natural limits were overcome, not because they were truly superseded, but most often because science and technology found ways to tap into natural pathways and potentialities so that the resilience of natural systems could handle the burden human extraction and excretion were placing on them. Yet, even so amended, the story of adaptation may be at a point of radical discontinuity. Can we reengineer the global atmosphere, or is geophysical fluid dynamics at last saying to us: "Enough is enough, change your way of living or reap the whirlwind"?

The discourse of accommodation and its underlying narrative of discontinuity are not new. Fifty years ago prescient economists like Kenneth Boulding were formulating the challenge we face today. "In the imagination of those who are sensitive to the realities of our era," Boulding wrote in 1965, "the earth has become a space ship, and this, perhaps, is the most important single fact of our day. . . . Man is finally going to have to face the fact that he is a biological system living in an ecological system, and that his survival power is going to depend on his developing symbiotic relationships of a closed cycle character with all the other elements and populations of the world of ecological systems."[2]

A Relational Turn

An ecological social contract will carry with it a different orientation and perspective on the political economy of our time. And, of course, it will change our way of thinking about what an economy (understood as an organized set of rules, roles, and relationships for production, distribution, and exchange) is for and how it should function. Within the framework of the consumptive social contract, the main disagreement has been between those who emphasize efficiency and those who emphasize equity; or between those who stress growth and those who stress just redistribution of existing wealth; or between those who own and control capital and those who own mainly their own labor and skills. These arguments are not passé, by any means; the struggle for fairness and equality has not been won.[3]

But an ecological social contract will be a framework in which an ontological reorientation will be added to an ethical reorientation. Ecological economist Peter Victor formulates this ontological reorientation in the following terms:

> One definition is that an economy is "the system of production and distribution and consumption." . . . A different conception of an economy . . . is as an 'open system' with biophysical dimensions. An open system is any complex arrangement that maintains itself through an inflow and outflow of energy and material from and to its environment. . . . Ecosystems are open systems. . . . Planet Earth is a closed system, or virtually so. A closed

system exchanges energy with its environment but not material. . . . And here is the rub. Economies are open systems but they exist within and depend upon planet Earth which is a closed system. All of the materials used by economies come from the planet and end up as wastes disposed of back in the environment. . . . As a result of [the physical laws of thermodynamics], open systems that depend upon their environment for material and energy must keep going back for more and must keep finding places to deposit their wastes. . . . Natural systems can be very effective in breaking down many of the wastes produced by people and machines, but often local environments are overloaded causing polluted land, water and air . . . [T]he scale and complexity of environmental problems have increased. Now we are confronted by broad regional problems . . . and global problems.[4]

An ecological political economy would be premised on the fact that the planetary systems that support life in its most fundamental physical, chemical, and organic manifestations have boundaries, tolerances, and thresholds.[5] These boundaries should—and ultimately will—constrain the extractive and the excretory activity of human individuals and societies. An ecological political economy would adopt a thermodynamic perspective having far-reaching policy and governance implications: human economic activity must be viewed not primarily in terms of its consumptive outputs—the satisfaction of human wants—but in terms of its systemic impacts—the transfers of both energy and matter that it brings about, in the context of thermodynamic and biophysical and chemical tolerances of the planet, which is a closed material system. The form of a fixed body of matter can be changed, but this should be done in a way that recycles its potential for a useful contribution to the ecology of life rather than degrading it into toxic forms. And the manipulation of matter involves both the production and the release of energy, which must remain within certain "Goldilocks" tolerances of atmospheric and oceano-graphic heating and cooling.[6]

A vast body of knowledge now exists that attests to the importance of respecting planetary system boundaries and thresholds. It is essential that these constraints be recognized and obeyed. Biological and ecological systems can process the material waste and excess energy produced by economic activity, but only within certain scales and tolerances. As plane-tary boundaries are approached (or exceeded), ecosystem functions are undermined and overwhelmed, thereby render-ing them—and the social systems that depend upon them—less able to support either human or natural communities that are flourishing and healthy, diverse, and resilient. No longer are only justice and dignity at stake; now minimally decent survival is in question as well.

What Does Nature Ask of the Human?

As I noted earlier, any social contract is a way of interpreting what human convictions should be and where human con-tentment can be achieved. Thus by placing economic activity in a natural context, an ecological social contract will also offer a different way of understanding normative or ethical goals and aspirations appropriate to conviction and content-ment. From the perspective of an ecological social contract, the accelerating, global extractive assault on planetary resources and ecosystems, as well as the unprecedented extensions of our technologic reach, do not truly represent progress and the triumph of human freedom or the human destiny. Why not? For one thing, as is by now familiar, they are not sustainable or viable as a road to the future, but must be bridled by the requirements of renewable energy sources. We have been drawing down the bank account of prehistoric photosynthesis for about as long as we can. Not primarily because the supply of fossil carbon is running out, but because the planet can no longer bear the systemic effects of its extraction and excretion. No less important, but less often noted, is the fact that the extractive and excretory assault of our economic behavior, and the technology that makes that behavior possible, contains an inner contradiction. While seeming to extend human freedom, economic behavior and technology are laying the groundwork for its repression; while seemingly representing the advanced expression of human capability, they are actually undermining what is most precious in humanness.

To find a healed relationship between humans and nature, how then should we think about humanness? On the reading of an ecological social contract that I offer in this book, the following two ideas are especially important in answering this question. First, there are the intertwined notions of innovative human agency and developmental capability: humans are remarkably good at doing new things, and they can improve or get better at what they do. Second, there is the moral imperative to treat the individual human being as a person—it is of great ethical value to be (and to be allowed to be) an acting, doing subject, rather than merely living as an object that is acted upon and done to. In short, humans are able to comprehend themselves as beings who become, as purposive agents who can live—and should live—without external domination. In his great essay on the concept of liberty, the political philosopher Isaiah Berlin wrote:

I wish my life and decisions to depend on myself, not on external forces of whatever kind. I wish to be the instrument of my own, not of other men's, acts of will. I wish to be a subject, not an object; to be moved by reasons, by conscious purposes, which are my own, not by causes which affect me, as it were, from outside. I wish to be somebody, not nobody; a doer—deciding, not being decided for, self-directed and not acted upon by external nature or by other men as if I were a thing, or an animal, or a slave incapable of playing a human role, that is, of conceiving goals and policies of my own and realizing them.[7]

In *The Human Condition*, another noted twentieth-century political theorist, Hannah Arendt, developed an anatomy of our humanness along these lines, using a suggestive but rather idiosyncratic terminology. According to Arendt, human beings are (1) creatures of "labor" who are subject to the biological rhythms of their organic needs; (2) practitioners of "work" who are subject to the creative encounter between natural materials and imaginative form; and (3) performers of "action," especially speech acts or communicative acts, through participation in the deliberative process of shaping common meanings in the public, symbolic order.[8]

The failure to live within planetary boundaries and limits—thereby turning our back on our interdependence with the earth and our own earthly, creaturely condition—will fundamentally threaten and transform the dimensions of labor, work, and action. Labor will produce illness rather than health. Creative work will become increasingly unavailable and unavailing. Action will devolve into bargaining and positioning for strategic advantage. At the dawn of the twenty-first century, precisely this baleful transformation in the human condition, this hobbling of human possibility, seems well advanced.

So we would do well not to underestimate the task facing us in reconstructing a social contract for the Anthropocene age and in moving from the consumptive to an ecological social contract. It is both a reorientation toward the natural world and an ethical innovation. It goes beyond the physical and life sciences in a descriptive sense and implicates the normative foundations of social order and human agency. An ecological social contract is a new story and a new conceptual framework within which we must make public policy and reform in the major structures of our society. It is the narrative of a journey of discovery concerning how to imagine, construct, comprehend, and govern a new form of social order that will achieve justice and empower flourishing life and living.

An ecological social contract will perhaps be able to offer a framework for the formation of democratic consensus through participatory deliberation. Grounded in solid natural and social scientific knowledge, as it ideally should be, an ecological social contract is emergent and dynamic, not dogmatic. The contract takes shape in action over time; it is not imposed by fiat. All economies appropriate natural matter and energy and, through labor, transform them into products for human use and exchange. But about this the consumptive social contract and its political economy are ecologically illiterate. To correct this, the coordination and organization of very large numbers of people, a vast massing of human agency, will be required in agriculture, mining and manufacturing, science and technology, transportation, construction, and the like. Such coordination requires a sense of common purpose, and the reconstruction of an ecological social contract is essential in the imagination and discovery of what that purpose should be.

A social contract holds a moral mirror up to each one of us. It is not only a narrative of discovering a new form of social order, it is also a narrative of discovering a new self-identity and a new way to live. In the market-oriented consumptive social contract dominant in the world today, the individual must live out the biography of the consumptive self, *homo economicus*: To survive and flourish, the economic self must fulfill (biological and psychological) needs. To meet one's needs, one must compete successfully to extract value from the labor of others or to secure access to positions in which one's own labor can provide the necessary income. To compete, one must understand and come to dominate the natural and social systems one inhabits. The consumptive social contract assumes the desire to acquire and consume to be psychologically unlimited. The individual then is compelled by his or her inner nature and external circumstance to appropriate and strive to dominate both the social and the natural environment. As a result, growth in the activities of extraction and excretion knows no bounds and perforce overcomes all other considerations.

In contrast, the kind of self called forth in an ecological social contract and its political economy has a quite different story line. To survive and flourish as an ecological self, you must fulfill (biological and psychological) needs. To meet your needs, you must compete and cooperate successfully with others to link your work (again in Arendt's sense) to that of others in respectful and accommodating designs. These designs will take advantage of the value and energy produced and reproduced by cyclic geophysical resources and counter-entropic living systems. Social arrangements must be such that you will receive just sustenance and provision in return for your work. In this way your needs will be met, and you will have the wherewithal to develop and pursue multiple human capabilities and courses of personal self-realization. In this narrative of self-hood, the desire to acquire and consume is not taken to be biologically impelled, psychologically imperative, or ethically unlimited. The individual then is not compelled by inner nature and external circumstance to strive to dominate either the social or the natural environment. As a result, growth in the activities of economic extraction and excretion will be bridled. These economic activities and the technologies that support them will be much more efficiently and intelligently designed to accommodate and work with nature, not against it. And they will be kept within sustainable ecological operating margins and normatively reasonable bounds. To coordinate your work with others in such a social arrangement (an ecological political economy), you must understand, care for, and respect the natural systems in which you reside, and you must critically and constructively participate in the cultural and social systems you inhabit.

How can we regain contact with the dimension of the human condition that Arendt calls "labor," namely our fundamental being as biological creatures, subject to the rhythms and requirements of life; how, in short, can we reconnect with our own animality and materiality, which we seem to have forgotten? How too can we rediscover our humanity in "work"? How may *homo economicus* (the self as gaming, calculating maximizer of personal utility) be transformed into *homo faber* (the self as craftsperson, responsible for and respectful of his or her materials)? Finally, how can we regain contact with "action" as the expression of our civic capabilities? How can privatized selves, trained to think only of a politics of "who gets what, when, and how," be nurtured so as to become discursive and deliberative democratic citizens attentive to the common good and to the obligations of trusteeship for the natural world?

An Enabling Act of Mind

We don't quite know yet how to foster the psychosocial development of such selves or write their collective biography on a large scale. I think it is undeniable that to some extent the terms of a new ecological social contract will demand that individuals and corporate entities submit to a regime of limits and common rules of restraint that are not compatible with the life narrative of *homo economicus*. In practice this means that the story line of political economy emanating from an ecological social contract must have recourse to values and purposes that the members of these societies will understand if they think and act like interdependent and relational selves— "ecological selves" and "ecological citizens." Part of the task of a new ecological social contract is to commit an enabling act of mind and imagination that can see the good and freedom in relational terms. The social contract of consumption over the years has helped to build a population of possessive individualists through its doctrines and through the institutions it has legitimated. Now we must be no less intentional about the task of educating a new generation of social persons. We must begin to fashion a pact of ecologically sustainable life—an ecology of the mind and of the heart—that will deserve principled conviction and foster genuine contentment and flourishing.

Part II

Natural Being, Cultural Becoming: Nature in Humans

Reason engenders *amour propre* and reflection fortifies it; reason turns man back upon himself, it separates him from all that bothers and afflicts him. Philosophy isolates him; because of it he says in secret, at the sight of a suffering man: Perish if you will, I am safe. . . . It is very certain, therefore, that pity is a natural sentiment which, moderating in each individual the activity of love of oneself, contributes to the mutual preservation of the entire species. . . . The human race would have perished long ago if its preservation had depended only on the reasonings of its members.

—Jean-Jacques Rousseau[1]

Democracy is a personal way of individual life; . . . it signifies the possession and continual use of certain attitudes, forming personal character and determining desire and purpose in all the relations of life. The task of democracy is forever that of creation of a freer and more humane experience in which all share and to which all contribute.

—John Dewey[2]

3

The Roots and Logic of Social Contract Theory

In this chapter I explore the roots and logic of the notion of the social contract, the central agreement out of which free and equal human beings create common rules and constraints that allow them to flourish together in ways they could not flourish alone. The idea of the social contract, regardless of how one fills out its substantive claims and content, is first and foremost a way of looking at the world. It is a conceptual framework that both enables and impedes the expression of deep insights and understandings. I turn to study the conceptual structure of early social contract theory because in it one finds a rich, complex, and nuanced understanding of three master categories of modern thought: nature, culture, and humanity or humanness. During the past three centuries, the idea of the social contract has linked the concept of nature or what is "natural" with the notions of culture and society. Nature helps us understand why culture and society exist and how their existence can be morally and politically justified. Moreover, social contract thinking aims to tell us what is distinctive about the humanity of human beings.[1] These are perennial concerns and it is no wonder that the social contract remains a meaningful mode of discourse today: it frames how we make sense of ourselves, our society and social relationships (especially our economic and political relationships), and the material and biological world around us.

The three greatest social contract writers are Thomas Hobbes, John Locke, and Jean-Jacques Rousseau. For this discussion I shall focus on Hobbes and Rousseau. I find their work particularly pertinent because they set out with clarity and power the basic political directions that the idea of the social contract can take.[2]

When the English Civil War broke out in the 1640s, it was both a religious conflict (Protestants vs. Catholics) and a power shift from an absolute monarchy to an early form of representative democracy (with a very limited franchise, but even then much more radical notions of democracy were in the air). The Puritan and parliamentary forces wanted to establish a republic, although it devolved into a military dictatorship under Oliver Cromwell soon enough. The Royalists were pushed back and the king's court went into exile in France. The revolutionaries eventually prevailed and King Charles I was defeated, tried, and executed in 1649. Well before that a leading figure in the ongoing political and philosophical debates, Thomas Hobbes (1588–1679), had also fled to France, where he was associated with the court, and the Royalists for a time promoted his work. Soon the supporters of the monarchy began to understand that his arguments could just as effectively justify the rule of Cromwell as that of the current claimant to the throne, Charles II. Hobbes fell out of favor with the court in exile and had to return to England in 1651, appropriately enough but somewhat ironically, under the protection of the parliamentarian government.

In his various political and psychological writings, Hobbes presented several versions of his account of the state of nature, the founding social contract, and the nature and justification for sovereign authority established by that collective agreement. The last and definitive version of his theory is found in his masterpiece, *Leviathan*.[3]

He tried to develop his political and ethical theory in quasi-geometric fashion, beginning from premises and propositions about human nature and psychology and drawing conclusions concerning politics and obligation. Human psychology for Hobbes was pain-avoiding and pleasure-seeking; human behavior—whether natural or cultural—was driven by the passions, not by reason. No ideas or moral virtues were innate. The main "natural" psychological motivation was fear, especially fear of violent death. What Hobbes called the state of nature or "the natural condition of mankind" was also a state of potential and actual perpetual conflict, a war of all against all. It was a condition before the invention of morality, with only one principle of natural right: self-preservation. That gave each individual unlimited freedom and license to do anything necessary to protect himself, including engaging in preemptive first strikes.

Hobbes is often mistakenly portrayed as holding a disparaging, bestial view of human nature. In fact, Hobbes did not posit that humans were naturally aggressive, cruel, or prone to violence. On the contrary, he thought that these behaviors were artificially constructed and that people were miserable in this condition. The state of nature/war was the result of insecurity, the lack of a creditable deterrent, the lack of authoritative common rules for mutual cooperation and living, and the lack of a common power to enforce those rules. In a word, the lack of a sovereign authority.

The social contract is the vehicle for creating that common sovereignty via consent of the subjects. Once established, the reasonable and legitimate authority of the sovereign is unified and absolute. Hobbes rejected the notion of constitutional government or separation of powers. There is no moral limit on the sovereign's laws. There is, however, one natural limit, and that is the right of self-preservation. If the sovereign threatens a subject's life, then the two revert to a state of nature vis-à-vis one another. The civic bond is broken, and with it all obligation, not so much because a higher natural right is invoked by the subject as because the sovereign's edict seriously threatening the life of the subject is a logical contradiction. It undermines the very grounds of sovereign authority—namely, protection and security of life. Survival of the strongest or most fortunate will be the amoral rule. Aside from the threat of impending capital punishment (and perhaps military conscription), however, the subject had no just cause for disobedience.

The Hobbesian sovereign defines the content of law and morality. Hobbes does introduce a notion of pre-political "natural law," but he treats these laws as prudent precepts, terms of a social peace treaty, not as rules demanded by reason at all times and all places, such as we now think of universal rights as being. Unlike Jefferson, Hobbes was not content to view natural rights as self-evident truths. Reason does not have that kind of power, and nature does not implant necessary moral laws within our minds and hearts, let alone within our genes and brains. Above all, the last thing Hobbes wanted to do was set up a standard of universal morality or religion outside the sovereign's law that would justify (and motivate) disobedience or collective revolution against state authority.

The positive laws of the sovereign need not be purely arbitrary, however. Strategic rationality and intelligence should guide the sovereign in promoting the commonwealth and in keeping the peace. Peace, tranquility, and order were the principal values in Hobbes's vision of civil society. Scientific knowledge should be promoted, but religious and metaphysical discourse should be curbed. Hobbes believed that it was religious extremism and the struggle to seize power under the guise of religious reform that led to the political turmoil in

England during this period, and the state of nature is an analytic portrait of the social and political chaos of his time. People living without sovereign authority are too self-interested to cooperate effectively because they have no reasonable grounds for reciprocal trust. Therefore, economic and technological progress suffers, the material standard of living declines, and the collective exploitation of natural resources is undermined.

During his lifetime, Hobbes had little direct influence on English politics. His defense of monarchy was not as reliable as the traditional arguments appealing to the divine right of kings that were deployed by thinkers such as Hobbes's contemporary, Robert Filmer. Indeed, social contract theory, as it was developed by Locke and then a century later by Rousseau, was the main ideological opponent of divine right theory in the modern period. And as mentioned earlier, although Hobbes personally supported monarchy, his theory did not necessarily or exclusively do so. Absolute unified sovereignty is what his theory establishes, and what Hobbes rejects is the regime of divided or shared sovereignty that historically had been associated with a "republic" or a "mixed constitution."[4] Unified sovereignty could take the institutional form of an aristocracy or even a democracy. Indeed, his account suggests that at the initial moment of the social contract founding the state, the sovereign power is in all the people, it is a participatory democracy. But he thinks that each individual will immediately transfer that democratic sovereignty to a unitary body (a charismatic leader or a small ruling elite) in exchange for protection, security, order, and what he nicely terms "commodious living." They will trade their natural equality and the democratic citizenship that exists for an instant in the first founding of the social contract for economic growth and material affluence. *The social contract of consumption is born.*

Jean-Jacques Rousseau (1712–78) also uses the social contract device to argue for the importance of unified sovereign authority. However, he develops a democratic version of the social contract in which the people collectively never do transfer their participatory democratic sovereignty, but instead reaffirm and reenact it periodically to keep those elites who undertake the day-to-day business of governance in check.[5] Moreover, Rousseau has a quite different understanding from Hobbes of nature in humans. In his account of the state of nature, conflict arises with the development of culture and the psychological transformation of individuals. The sorry state of humankind that Rousseau saw in his day, and would surely see in our day if he were alive, is a contingent cultural and historical condition, not a natural or inevitable one.

Nonetheless, in many ways, Rousseau's thinking about the state of nature and the social contract resembles that of Hobbes more so than it resembles Locke's, but as a latecomer to the traditional discourse of social contract theory he could critically position himself in relation to each of his predecessors. Rousseau believed that Hobbes had improperly projected conflict that is the product of *social* existence onto *natural* existence, which he conceived as lacking in sociality. Likewise, he thought Locke projected reason and a form of morality (external natural laws and the inner capacity to understand and obey them) mistakenly onto the state of nature, when what he was really describing was a state of society of a certain kind. Rousseau was determined to reach back to bare nature, but what he found there was essentially human absence. Both conflict and orderly moral and social living were part of a narrative of movement from the natural state to the social state—each are inherent possibilities of humanness, but neither are inevitable expressions of it.

In his main work of political philosophy, an incomplete and rather cryptic book entitled *On the Social Contract* (*Du contrat social*), he presents his conception of the motivational and moral basis of sovereign authority in a human political culture. Like Hobbes, he probes deeply into the psychological and ontological sources of sovereign power as the only alternative to insecurity, conflict, mean living, and death. For his part, Locke anticipates the later work of theorists in the liberal tradition in that his main concern was freedom, fairness, and impartial justice. Hobbes and Rousseau played for more basic stakes: life itself and the natural and social conditions of life's flourishing.

Hobbes saw sovereignty embedded in political institutions—monarchy, aristocracy, or democracy—and the establishment of sovereign authority grew out of a reciprocal agreement, every man with every man, to have their private wills represented by and become subordinate to the will of the sovereign representative. Rousseau saw sovereignty as what he called the "general will" (*la volunté général*). This is not a person or an office; it is a moral culture and a spirit of community and solidarity. It is a capacity for a radically demanding impartiality of moral intention and the scrupulously public use of political power. The general will stands behind all other forms of legitimate rule and power in a society. It is socially and politically embodied in a participatory republic, that is, in a set of interlocking governance institutions that rest on the periodic re-creation of the general will by a popular assembly of all citizens. And the general will is culturally embedded in a special form of moral imagination and civic virtue, one informed by the independent judgment of each citizen and by a strong, widespread sense of common need, vulnerability, and equal concern and respect. The Rousseauian social contract creating the sovereignty of the general will is a rare, difficult, and precarious achievement in human history (has it ever truly been realized?), but it is the highest expression of our communal humanity, and it arises from *the development of humanity out of nature through culture.*

Unlike Hobbes in *Leviathan* and Locke in the *Two Treatises of Government*, Rousseau does not present all the elements of his version of social contract theory in a single work. In *On the Social Contract*, he tries to give the contractarian dialectic of nature and culture a kind of rigorous, logical form. There he delights in showing the inconsistencies and false inferences of other modes of theory and political reasoning. His interesting and original account of the state of nature is cast as a conjecture about the phylogenetic development of humanness and the origins of social inequality, and it is set out most fully in his *Discourse on the Origins and Foundations of Inequality among Men.* No less essential to Rousseau's understanding of nature and culture is *Émile*, a hybrid work, part novel and part educational manual. The developmental trajectory that the *Discourse* examines on the phylogenetic or species level is complemented by a parallel developmental account on the ontogenetic or individual level. It traces the developmental and dynamic formation of Émile's sense of self and conscience, from infancy to adulthood, from amoral natural animal to virtuous natural man, and perhaps ultimately to civically virtuous citizen—if only he could ever find a republic worthy of his loyalty and participation.

In these and other works, Rousseau showed not only the logical acumen on display in *On the Social Contract*, but a remarkably critical eye as a social and cultural observer of the society of the last years of the French *ancien régime.*[6] The central concept of his cultural and social criticism was *amour propre*, literally "self-love," a term impossible to define precisely or simply, but which carries a cluster of meanings, including vanity, self-regard, the tendency to get one's sense of self from being seen and approved of by others, insincerity, and inauthenticity. As *amour propre* came to dominate the identity and motivation of modern man, it became the driving force behind injustice, conflict, and the eclipse of true moral individuality and citizenship, and also behind the security-providing power of Hobbesian sovereignty and subjection. *Amour propre* lies at the heart of the social contract of consumption. It must be supplanted by other forms of conviction and contentment in an ecological social contract.

A Logic of Nature and Culture

The social contract can be understood as a pact of self-interested restraint based on fear and the desire for competitive advantage (Hobbes) or it can be a condition of equal membership and caring mutuality based on a sense of duty toward oneself and others to pursue and realize the common good (Rousseau). What is the root of these two alternative social contracts? How do the natural and the cultural interpenetrate and combine to produce the conditions that make humanity possible—that make us most fully human? In addressing these questions neither Hobbes nor Rousseau took human nature or, as I would prefer to say, nature in humans, to be the limiting essence determining the scope of our human possibilities. Unlike many contemporary thinkers in the social sciences and the humanities, they did not see culture as merely a veneer overlaying nature, nor did they see nature as a mere figment of cultural imagination, a construct. Instead, their account was more complex and dialectical. Nature and culture were not really distinct, but ontologically cocreative. Nature alone cannot realize humanity. The divine is no longer on stage, nor is it an important part of the story the early social contract writers tell.[7] Culture brings forth human self-realization and the human good—on both the individual and the communal level. But culture can do this only by reaching into and drawing on nature. The cocreation of humanity can go wrong. Culture can also deform and corrupt; it can actualize the human stain as well as the human good. In either case, culture acts in concert with nature to produce some form of humanity. In an important passage Rousseau adumbrates the possibilities of a perspective that moves away from natural or ontological essentialism, although we must bear in mind that it is only one of several formulations Rousseau offers on this recurring theme— indeed, the master theme—of his work. "The passage from the state of nature to the civil state produces a remarkable change in man," Rousseau writes:

> by substituting justice for instinct in his behavior
> and giving his actions the morality they previously
> lacked. . . . Although in this state he deprives himself of
> several advantages given him by nature, he gains such
> great ones, . . . that if the abuses of this new condition
> did not often degrade him beneath the condition he left,
> he ought ceaselessly to bless the happy moment that tore
> him away from it forever, and that changed him from a
> stupid, limited animal into an intelligent being and
> a man.[8]

Goodness and justice are culturally natural—nature's potential realized through culture. Yet evil and injustice are also "natural" in their potential and cultural in their realization, their embodiment in social action, and their embeddedness in historical time. Morally and politically Janus-faced, nature can be well realized through culture into virtuous, communal human life, or it can be deformed through culture into vicious, deracinated human life.

Why not simply say that nature is value-neutral and that both good and evil, justice and injustice, virtue and vice come from culture? Because the concept of culture alone cannot bear that philosophical or anthropological burden. The concept of culture must operate in an interplay with the concept of the natural, by which I mean both the infrastructure and the potential of human being, its *bios* and its *telos*. Only then do the form-giving, rule-governed, meaning-making capacities of culture become sources of normative force and interpretive understanding.

In its argumentative structure and logic, social contract theory explicitly rests on a contrast between a cultural and a natural form of life or condition of existence. The natural condition, which is commonly called the state of nature, is used both as a vehicle for presenting the theory's conception of what is rooted in (human) nature and as a heuristic construct that clarifies the essential features of the cultural condition, both ideal cultural conditions and deformed, unjust ones. The state of nature is mainly interesting for what it lacks and as a way of instructing us about the human consequences of that lack.[9] What it lacks, of course, are precisely those features that at least partially define an ideal, teleological notion of natural and cultural becoming, both phylogenetically, in the development of the human species, and ontogenetically, in the development of each individual person. In other words, for social contract theorists those elements that are said to be absent in nature are carefully selected to reinforce and illustrate the theorist's view of what is most important about cultural order. Nature need not be something prior to or outside culture; it can be an imminent ideal within culture, a critical standard against which an actually existing culture or political system can be evaluated.

Thus far I have made reference to the concept of culture several times, but it is a complex and protean notion in social science and social theory. Many social scientists, especially anthropologists, today tend to understand culture as a general symbolic order and framework of meaning.[10] Clifford Geertz elegantly expressed the gist of this approach: "Man is an animal suspended in webs of significance he himself has spun; I take culture to be those webs, and the analysis of it to be, therefore, not an experimental science in search of law, but an interpretive one in search of meaning."[11] In a similar vein, David Korten has more recently formulated aspects of this perspective that are pertinent for my concerns in this book:

> Culture is the system of beliefs, values, perceptions, and social relations that encodes the shared learning of a particular human group essential to individual survival and orderly social function. It serves as the interpretive lens through which the human brain processes the massive flow of data from our senses to distinguish the significant from the inconsequential, assign meaning, and shape our behavior.[12]

Contractarian political theorists, such as Hobbes and Rousseau, already had a general appreciation of this system of beliefs and meanings, but their concerns were somewhat more focused. For them the defining features of "the cultural" are the systematic and institutionalized organization of power and authority, the system of orderly coordination and cooperation among large numbers of individuals, and the sense of membership and common interest. In short, the social contract writers like Hobbes and Rousseau start with something they call the state of nature, or the original condition, characterized by an absence of form—institutional form and psychological or motivational form. *Homo sapiens* is a creature propelled by drives and reasons, but not much by instinct; a creature with a remarkably wide behavioral repertoire and an extraordinarily flexible adaptive capacity.[13] Social contract theory then goes on to tell a story about how fluid natural being is turned into rule-governed cultural becoming. That is to say, it tells us how a place is to be created (civil society) in which the potential creatures we are can become actual and fully human, however that is to be characterized—rational, cooperative, productive, moral. The moment of transition, in the social contract story at least, is an act of promising.

Consider what it means to promise and what it means to be able to promise. It is an act of self-binding. It is an act dependent upon a conception of time; unlike contemporaneous barter or exchange, promising is an act in the present that imagines action in the future. When the time of fulfillment arrives, the promise then binds the present to the past. It requires the capacity to communicate in a form that allows the complex function of tense. Moreover, the act of promising has force only if there is some stability in the identity of the selves involved over time. Strictly speaking, the act of promising does not require a prior relationship between the parties, but if you add the element of trust that is most likely to be necessary for a promise made to be accepted as such, then promising is a practice that makes sense only in the context of some kind of ongoing association. As demanding as the act of dyadic promising or small group promising is, how much more so is the notion of an act of collective promising involving an entire society?

That is surely one reason why few (if any) social contract theorists regard it as a historical event. If one supposes that the social contract is the moment of transition from a time when human beings were not capable of promising to a moment when they suddenly have this capability, then the notion makes no sense. Rather the notion of the promise as the touchstone of our humanity, both enabling and constituting our human potential, must be seen as a permanent possibility, always before us but regularly slipping beyond our reach and eluding our grasp. The social contract is the interdependent and reciprocal promise by each to all and by all to each that is rarely ceremonially reenacted but that is made, or broken and betrayed, through our actions from moment to moment in our everyday lives.

In Hobbes and Rousseau, the promise of the social contract is looked at from the perspective of before and after, even though as I have just indicated, that distinction is problematic. Before the promise is the state of first nature; after the promise is the state of second nature or political society. This move from before to after is a structural as well as a temporal one. It takes humanness from a condition of moral and social isolation (Rousseau turned this into true solitude, Hobbes and Locke did not) to sociality and interdependence.

Rousseau does something novel in that he offers two very different accounts of the contractarian moment. In the *Discourse on Inequality* he offers a hypothetical phylogenetic narrative (an evolutionary cultural and psychological history of the species) culminating in a phony social contract in which the wealthy few dupe the simple many and extract collective consent from them.[14] The duplicitous social contract is built on corrupt foundations. The people are susceptible to this trick because they have lived during what he calls the "Golden Age," an already rich and complex cultural order that contains the elements of transparency, justice, love, community, and empathy he is looking for in the authentic social contract.[15] The authentic contract, outlined abstractly in *On the Social*

Contract and summarized in *Émile*, consolidates and protects the human virtues of the Golden Age, much as the phony contract poisons and eventually transmutes them into pride, vanity, greed, the will to power, duplicity, insincerity, and loss of true self.

However the theorist may flesh out and manipulate nature and culture, a common element in social contract theory is its critical power. This critical power has three aspects.

First, as has been noted, social contract theory motivates the human story through a critique of the absence of form in its conceptualization of nature. The raw material of humanness is there, but the shape it can take has yet to be determined.

Second, it presents an ideal conception of human affirmation in a just cultural order that is founded by the social contract.

Finally, it mounts a critique of presently existing social and political order by way of contrast—the existing order does not remedy the deficiencies of nature, or else it distorts nature's possibilities. The existing order fails to embody the ideals and purposes animating the aboriginal founding, and thus it fundamentally lacks legitimacy and authority. It rests on a deception and a kind of misdirection of the human story.

When a social contract theory lays out what it takes to be the appropriate general form of cultural being, it thereby establishes an evaluative standard, a regulative ideal, against which any particular existing cultural arrangements can be judged.

4

The Uses of Nature and Culture

Artifice and Accommodation

In the previous chapter I concentrated on what might be called the "outer" order of humanity—social, cultural, and political order. The state of nature is a philosophical device used to establish both the necessity of such order and the justification of a particular form that this outer order can take. Social contract theory can also develop an analogous critical perspective on the "inner" order of humanity—psychological, motivational, aesthetic, and intellectual order. Again, in regard to inner form, the state of nature is philosophically illuminating by dint of what it lacks, and its use as a negative heuristic device, so to speak, can be applied to the theory's underlying conception of well-realized humanity or the human good. When it is so applied, social contract theory provides a powerful conception of humanized nature or "second nature," which is the natural cultivated. I say cultivated, not subdued, replaced, or turned into sheer artifice, because I am struck by the deliberateness of the contractarian thinkers in seeing the cultural state of humanness as *accommodating and building on* natural tendencies and potentialities, which nature alone is unable to bring to fruition in human beings, not as *creating and controlling* those tendencies and potentialities out of nothing.

Nonetheless, within this impulse to build on rather than to replace nature in order to achieve humanity (that is to say, in order to achieve a cultural ecology in which humanity can flourish), two significantly different emphases are possible.

This difference, in my estimation, is one of the most important things separating Hobbes and Rousseau. For Hobbes, to move from the state of nature into cultural order is to change the behavioral manifestations of the natural in humankind by changing the interactive environment (the lifeworld) and the strategic logic (the system) in which that nature expresses itself. For Rousseau, the transition from nature to culture is a qualitative transformation of the natural.

Let me try to formulate this in a slightly different way. For Hobbes, the inner order of humanity was complete and given already in nature—only the outer, cultural order of properly realized humanity was naturally absent. Humans lurking in ambush in the state of nature (i.e., the war of all against all) are not fundamentally different creatures from what they are when they walk securely and confidently in the streets of Leviathan state.

For Rousseau, by contrast, humanity is in ontological motion. This applies to human beings both as they (potentially) exist in the state of nature and as they (actually) exist in civil society. The human lies at the intersection of the natural and the cultural, and it is dynamic and perfectible. Right cultural order is to be evaluated both by the plenitude it offers in the face of nature's absences and by the extent to which it serves as a developmental medium for the evolution of inner nature. It is not altogether easy to say how this transformation should be characterized. Rousseau often resorts to paradoxical formulations, for example in *Émile*, where he is at pains to show that it takes a great deal of artifice to make man natural. Culture can build on and bring out the good in nature—it can create virtue and the moral will—just as it can also bring out the worst in humanity—a deformed and improperly realized humanness. But nature also brings out the good in culture. Neither concept points to a state of being sufficient and perfect unto itself. Only in their interaction do we find the good state of human being.

Especially for Rousseau, human nature was not an essence; it was not a set of timeless and universal properties that establish the parameters of cultural and moral possibility. Therefore, the becoming that takes us from nature to culture is a transformation that constitutes humanity. It can be a transformation in which, through culture, the distinctively human level of being is either elevated above nature or degraded beneath it. In a move anticipated, but not pushed as far by Hobbes, Rousseau recasts the traditional idea of a human nature or essence that is timeless and universal into human nature as the developmental emergence of humanity out of the cocreative or dialectical interplay of the natural and the cultural. Properly realized or well-formed humanity (marked by motivations of civic virtue and impartial will) grows out of the cultural and political structures he sketched in *On the Social Contract*. Improperly realized or deformed humanity marked by *amour propre* (an insecure and duplicitous identity, and strategically self-regarding and manipulative behavior) grew out of the social and technological features of modernity, although its roots lie as far back as the invention of private property, the environmental exploitations associated with agriculture and mining, and even with the origins of language itself, and hence the origins of self-consciousness and imagination.

Hobbes and Rousseau developed conceptions of the natural foundations of cultural order that were independent of any transcendent metaphysics, cosmology, or theology. They maintained that cultural order is the artificial, rational construction of human will. But they went even further and argued that in constructing cultural order, human beings have no transcendent pattern to follow. Without the benefit of transcendent reason or revelation, humankind becomes not simply the builder of cultural order but, in a much more radical sense, its architect or creator. The raw materials for this creation may be given, but the blueprints are not. Moreover, for Hobbes nature does not favor or dictate one particular or best institutional form of governance, provided that the sovereign authority is unified. In theory, all three of the classical regime forms—monarchy (rule of the one), aristocracy (rule of the best few), and democracy (rule of the many)—could fulfill this requirement. Although from a logical point of view his theory was indifferent as to the institutional form the unified sovereign may take, he preferred the form with (he thought) the greatest centripetal force, namely, monarchy.

Now, if humankind is the creator of cultural order in an ontological sense, the practical task facing particular men and women who must create and sustain—produce and reproduce—a particular cultural order requires that they too must look inward for a radical and exacting kind of self-knowledge. By the time he wrote *Leviathan* in the 1660s, Hobbes had come to think about sovereignty, not as vested in the one, the few, or the many concrete individuals, but in a more abstract way as vested in the complex practice of representation. The sovereign is what Hobbes called a "representative person," an individual (or group) who makes present in a useful form the united will or consent of the members of the body politic, and who embodies that will as an effective collective power. The exercise of this power requires knowledge of human nature, beliefs, passions, and interests, so that the real subjects of the sovereign's rule can be persuaded, predicted, and controlled. This is a task that is daunting even today, but all the more so in an age when the media and the metrics for assessing such information were not available. Yet even if he could have done so, Hobbes would not be interested in taking a poll. His solution to the challenge of finding a knowledge base for effective rule or governance was ingenious. It involved what might be called the inward turn of the political imagination. Conceived as a representative person, the Hobbesian sovereign needs only to be able to read himself, but in a particular and demanding way: to read in himself humankind and to read humankind in himself. *Leviathan* was written to teach him to do precisely this. But who could absorb its teaching? In order to succeed, *Leviathan* had to find only one suitable reader, but how much more difficult is that task than achieving sales in the millions?

To successfully ground his more participatory and democratic conception of the ideal cultural order, Rousseau had somehow to extend the diffusion of this culturally creative self-knowledge more widely than did Hobbes. The viability of the Rousseauian cultural order demands a more positive, active, and energetic mode of citizenship by a greater number of individual members of the cultural community than the Hobbesian cultural order demands for its viability.

Human nature—considered as a complex of rational capacities, passionate drives, and inherent potentialities—provides the starting point for the emergence of humanity as constituted by both inner psychic order and outer cultural order. But the possibilities of our nature will always exceed the actualities of our realized humanness. Thus, social contract theory, at least as Hobbes and Rousseau fashion it, has an element of tragedy. Whatever cultural condition of life is built on it, human beings will be marked by estrangement from nature. Even a political or civic humanity that is properly fashioned and realized in a virtuous and just culture will nonetheless, by Rousseau's lights, be a kind of "denatured" humanity. This view of ontology or philosophical anthropology has important implications for how we approach the problem of reconciling the conditions of flourishing natural life with those of flourishing human life. To achieve this reconciliation in a sustainable way, must humankind return to a closer symbiosis with nature or lessen our connection with, and destructive impact on, nature through technological mediation?

For Hobbes, that estrangement is endemic and ontological. Being is merely matter in motion; living being is metabolic in the sense that it has a basic impulse, the "endeavor" toward self-preservation. Hobbes's overall conception of the human condition is captured in this remarkable passage:

> To which end [i.e., those qualities of humankind that concern their living together in peace and unity] we are to consider that the felicity of this life consisteth not in the repose of a mind satisfied. For there is no such *Finis ultimus* (utmost aim) nor *Summum bonum* (greatest good) as is spoken of in the books of the old moral philosophers. Nor can a man any more live, whose desires are at an end, than he whose senses and imaginations are at a stand. Felicity is a continual progress of the desire, from one object to another, the attaining of the former being but the way to the latter. The cause whereof is that the object of man's desire is not to enjoy once only, and for one instant of time, but to assure forever the way of his future desire. . . . So that in the first place, I put for a general inclination of all mankind, a perpetual and restless desire of power after power, that ceaseth only in death. And the cause of this is not always that a man hopes for a more intensive delight than he has already attained to, or that he cannot be content with a moderate power, but because he cannot assure the power and means to live well which he hath present, without the acquisition of more.[1]

For Rousseau, the sources of this estrangement are both ontogenetic and phylogenetic. At the individual level, it resides in the betrayals of infancy and childhood, malicious (as a boy, Rousseau was locked outside the gate of Geneva by a malign gatekeeper who deliberately, Rousseau thought, closed them early) or inadvert (Rousseau's mother died in childbirth and thus he remarks, "my birth was the first of my misfortunes").[2] At the species level, it resides in the fatal rise of self-consciousness and imagination that is inseparable from the acquisition of language, and with it distinctions between self and other, mine and yours.

To be viable and sustainable, any cultural order must be compatible with the psychological possibilities determined by human nature. Of any conception of the social contract ask whether the psychology on which the theory is premised can plausibly reproduce over time the civic motivation that a just cultural order needs to sustain itself and its politics. Hobbesian natural humans become citizens by the push of fear and advantage. But surely citizens are needed most precisely when it is vital to overcome fear and advantage, when the push of self-interest must give way to the pull of moral ideals. Rousseauian natural humans become corrupt cultural humans by duplicity and by the triumph of *amour propre* over *pitié* (compassion, care, spontaneous aversion to suffering). Living in the age of Enlightenment, Rousseau believed that his corrupted contemporaries in France could not become citizens. But those in the backwaters or byways of modernity (like the Genevans or the Corsicans), or those specially tutored (like Émile), may become citizens (or at least become fit for citizenship should they find a just republic to join) by returning to the natural sources of morality anterior to reason in nature—*amour-de-soi* (care for the self) and *pitié*, but not *amour propre*—and reconstructing them in artificial form as general political will and civic virtue.

Moreover, for both Hobbes and Rousseau, nature is not constituted by reason; reason is a component of culture. For Hobbes, reason is merely the capacity to relate desired ends to efficient means; rational thoughts, as he put it, "are to the desires, as scouts, and spies, to range abroad, and find the way to the things desired."[3] For Rousseau, nature is constituted by love, and the basic story of the development of humanity through culture, human self-becoming, is the story of the appropriate or inappropriate cultural apprehension of natural love in cultural and political form.

This structure of ideas indicates how complex matters become when theorists recognize (as Rousseau did more adequately than Hobbes) that the politically pertinent psychological traits of human beings are themselves the products of both nature and culture and must be understood in light of both. For Hobbes, these psychological traits are essential and timeless, but they manifest themselves only contingently and culturally according to a logic of strategic patterns of social interaction. For Rousseau, the inner order of humanity is potentially given in nature but has two possible destinies: a virtuous communal and moral one or a corrupt one of competitive and possessive individualism.

5

Re-enchanting the
Social Contract

As we have seen, neither Hobbes nor Rousseau was a rational-
ist thinker. Hobbes preceded David Hume in regarding the
passions and emotions as the primary causes of most human
behavior, with reason playing at best an instrumental and strategic role.[1] Rousseau also was most concerned with those forms
of *amour* that were prior to the development of reason in
human evolution and were psychologically stronger and more
important in individual behavior.

Nonetheless, social contract theory, perhaps thanks in part
to Locke, has gained the reputation of being the ideology of the
commercial bourgeoisie. Contractual relationships and deal-
ings were displacing more traditional norms and customs as
capitalism replaced feudalism in early modern Europe.[2] Looking back on this massive social change, later historians saw the
idea of contract as a leitmotif of individualism and loss of community. The noted legal historian Sir Henry Maine organized
his account of such a transition around the shift from "status"
to "contract." Sociologist Ferdinand Tönnies wrote about the
shift from "community" (*Gemeinschaft*) to "society" (*Gesell-
schaft*), meaning essentially a social order in which cold, impersonal contractual relationships dominate. Intellectual historian
Benjamin Nelson studied the shift from what he called "tribal
brotherhood" to "universal otherhood." Political theorist C. B.
Macpherson linked social contract thought in Hobbes and
Locke with what he called "possessive individualism."[3]

This tradition of historical and social scientific discourse has been the backdrop to my account of the consumptive social contract as one substantive version of the social contract idea that is linked to the special characteristics of capitalism, especially the later capitalism of the fossil carbon fuel era that exploded in the nineteenth century.[4] But my own view is that social contract thinking is much more than merely an ideological rationalization for class interests. It can be that in some circumstances, of course. Even complex thinkers such as Hobbes, Locke, and Rousseau are engaged in specific struggles over policy and power. Their work is interpreted and appropriated by those with specific vested interests.[5] But the most searching thinkers also show that social contract thinking offers a framework that can accommodate critical insights and values of a more universal and enduring sort. Contractarian thinking is as contractarian thinking does.

In setting out an opposing substantive version of the social contract for a post–fossil fuel and ecologically oriented future, I need to consider elements of the transition to an ecological social contract and an ecological political economy that do not fit with the strong rationalism of this capitalist and commercial worldview. In order to develop the discussion further in this direction, I propose, again with Hobbes and Rousseau as touchstones, to devote this chapter to the distinction between what I call "enlightenment" and "enchantment." Enlightenment relies on empiricism, materialism, science, and reason. Enchantment relies on an interpretive sensibility and judgment. Where enlightenment examines a causal nexus in a reductive and instrumental way, enchantment explores webs of interconnection and meaning using narrative and figurative modes of understanding.

If what may be imperative now is a transvaluation of values at the level of both conviction and contentment, something that does not occur often but has punctuated our history from time to time, then the appeals of both enlightenment and enchantment will be important. If we have to change our *doing* in drastic ways, for the sake of ecological health and resilience as well as social justice, then I believe that we will also have to change our thinking about *being*. The historian of science and technology Robert Nadeau has addressed recent theoretical and research work in physics and cosmology, neurology, evolutionary psychology, and comparative religion. In science he finds the discovery of processes of ontological convergence and an underlying unity of being. In evolution and religion he finds the development of a cosmological imagination or sensibility that is both cross-cultural and transhistorical. Nadeau believes that in this kind of cosmological imagination, much more so than in secular materialism, we can find the touchstone to motivate and direct urgently needed social changes.[6]

Novus ordo seclorum

We are nearing the close of the current chapter of human history, which contains the story of a materialistic and secularized worldview. Out of this process grew Enlightenment culture and political liberalism—parents of the social contract of consumption. Earlier chapters have touched on the materialization and mechanization of nature. In this chapter I turn to the issue of secularization in a special sense. Secularization pertains to the loss of an imagination of transcendence from the warp and woof of our public culture and experience. Perhaps loss is too strong a word; the "privatization" of the transcendent is closer to the mark. This leads to the fact that an appeal to a context of being or experience or value larger than ourselves has lost its self-evident or straightforward intellectual attraction.

In one's private life of personal belief, religious practice, or spirituality, the language of transcendence still makes sense and still can be understood by others as a sincere and authentic way to articulate a meaning that people can share in common. Not so in the public domain, when this language is heard and interpreted, for the most part, ironically. It is code, a misdirection, or a hypocritical attempt to gain support. It is one thing that in secularized political discourse, one cannot successfully carry the day any longer by invoking the will of God. It is another that one cannot appeal even to a notion like the public interest or the common good without arousing suspicion. Utilitarian versions of ethics predominate at a time when claims about the intrinsic rightness or wrongness of some action or state of affairs become suspect and lose cultural legitimacy. Decisions then tend to become instrumental, calculating, and often solely materialistic. This tends to happen when appeal to a transcendent source of value or purpose is blocked off.

In a similar vein, it is virtually impossible to appeal against corporate behavior or market forces by invoking the notion of a "just price" or a fair share. Consider what a just price might mean from a substantive point of view. The dominant understanding of appropriate pricing is procedural: the price of a commodity is set by a market exchange involving supply and demand. If many people want something, they will be willing to pay a high price, especially if the commodity is scarce and not widely available for sale, and if there are not many sellers who might compete with one another by lowering the price they are asking.

Many central tenets of mainstream economics flow from this (admittedly simplified) ideal-type model of the market as a kind of self-regulating or cybernetic mechanism. If the price of something is high, new sellers (producers) are attracted into the marketplace, and competition among them will drive the market price down. Then as some producers drop out of the market for a given commodity, its supply will dwindle and the price will rise.[7] Also, if the demand for something is fulfilled, consumers will turn to something else and the price of the commodity will fall, and perhaps eventually the supply of that commodity will dwindle or disappear.

Undue concentrations of public or private power to control buying and selling activity—such as monopolies of supply or government regulation—distort this mechanism. So do extrinsic values that limit the individual pursuit of material self-interest. Some of these nonmarket values may be reprehensible, such as racism, and there is a purely economic argument against them. But other nonmarket values are laudable, such as social justice and equal dignity and respect. How can we avoid throwing the baby of duty out with the bathwater of discrimination?

Many years ago I attended a seminar led by Bayard Rustin, a prominent civil rights leader. At that time Rustin was at odds with more radical black power advocates, and was seeking to bring the African American community into mainstream entrepreneurialism and market society. Rustin favored attacking the problem of racial inequality and injustice in America as an economic problem first and foremost, rather than as a racial and cultural one. I recall vividly one thing that he said, and it bears on our discussion here. He argued that social justice required appealing to the economic desire for profit and to the consumerism of the majority white society so that their economic self-interest would pull them toward equal rights and treatment of minorities even if their discriminatory cultural and value attitudes pushed them away from social justice. Referring to the motivational power of financial self-interest, he said that the best hope for African American progress and equality lay in the fact that the white man "loves green more than he hates black." Today we need to pose a similar question: Can an ecological social contract count on a love of green money overcoming, if not a hatred, then at least a careless disregard for green nature?

Let us return to the idea of the market as a cybernetic self-adjusting mechanism that determines a just price. Two things are particularly noteworthy about this. The first is that it rests on the assumption of the continuous production of desire or consumptive demand and the continuous production of things to satisfy it. Commodities that are desired are relentlessly placed on offer by this market society. Producers will multiply in number so more is produced to meet profitable market demand or new technologies will be invented to increase production more efficiently. This is a conveyor belt of desire. If it clogs, slows, or breaks down—increasing unemployment, loss of consumer confidence, a liquidity crisis in the financial system after defaults or bankruptcies of large institutions—then the economy as a whole falters and the social contract of consumption is fundamentally challenged. A significant slowdown in economic activity offers the paradox of resulting ecological benefit at the cost of widespread social harm. (Atmospheric carbon levels went down for a time in the recession that followed the global financial crisis of 2008–9.)[8] A severe recession or depression is a violation of the terms of the consumptive contract that, if it persists very long, will trigger a democratic political reaction or a loss of legitimacy and credibility on the part of the state. On the other hand, human extraction and excretion on a planet that is a closed system in terms of matter, but an open system in terms of energy cannot be unlimited in its quantity and its pace. We can't do too much, too fast ecologically, and yet doing ever more, ever faster is precisely what the market system and consumptive society require economically and ultimately politically. This is the conundrum of the social contract of consumption.

The second thing to notice is that this idealized model of market efficiency is the current understanding of what might be called the "just" price. The just price is simply defined as what a properly functioning market exchange system will cybernetically determine. But this conception is not in fact purely procedural because it is grounded on one substantive value, namely, the value of fulfilling limitless individual human desire. Desire and demand for each separate commodity may wax and wane, but desire and consumer demand as such cannot be denied in such a conception. There then is no transcendent value or purpose, such as justice has in the past been thought to be. There is only the individual human fulfillment of desire through consumption and the relationality of competitive market exchange. The self-esteem of successfully competing—more efficiency if you are a producer, more savvy bargain hunting if you are a consumer—is everything. This is the heart of what Rousseau meant by *amour propre*, the serpent in the garden of the human story on both the species and the individual level, as far as he was concerned.

Consequently, we find ourselves attempting to govern in a conceptual environment hobbled by the exclusion of any ontological or ethical claims that are not human-centered. If we cannot take seriously a nature that has meaning or a sense of something sacred ontologically, then we also cannot take seriously social relationships and communal purposes larger than our own self-interest or the interests of our intimates and close associations. That is the reason the question of the role of the transcendent in a social contract is important. We must rethink the exclusion of the transcendent from the public realm.

And then, I think, we must find a way to integrate and synthesize enchantment (imagination and wonder) and enlightenment (reason and control) as complementary and mutually essential for a new ecological social contract and worldview.

Keeping Ontology Outside the Gates of the City

This is not an easy issue to broach or even to formulate without falling into positions that undermine right relationship and right recognition rather than support and reinforce them. Right relationship calls for restraint, duty, and responsibility, and it is tempting to turn to religious traditions and language for the transcendent meanings that readily support these virtues. But right recognition calls for inclusiveness, diversity, and plurality of voice and perspective. The quarrel between liberalism and religion has been over how to prioritize right relationship and right recognition—righteous living versus righteous toleration and diversity, one might say. Getting these priorities straight is key to the success of any social contract, but they are dynamic. Perhaps, as we will see, the greatest wisdom lies in insisting that there should always be a place for both in our public lives as citizens as well as in our private lives as congregants.

However this may be, historically fundamental to the liberal tradition is the exile of metaphysics, theology, and what have been called "strong ontological arguments" from social, economic, and political life. Liberalism and market societies have rejected strong ontological beliefs as an appropriate part of political or public discourse.[9] For example, John Rawls, the most recent doyen of social contract political theory, has argued that no "perfectionist" or thick theory of the good can be a part of the logic and discourse of the social contract.

For him that means that such substantive conceptions of the good (in which he includes religious belief) should play no role in deliberations leading up to an overlapping consensus on the principles of justice that should inform the institutional structure of a pluralistic society of free and equal persons. Nor should they play a role in the ongoing political discourse of public reason in the democratic affairs of a just society.[10]

Rawls is a very complex and sophisticated thinker, but more pedestrian versions of this suspicion of strong ontologies abound in contemporary diverse, multicultural societies. These suspicions are reinforced by cultural and religious revitalization movements in the world today, often spawning repressive revolutionary movements and illiberal theocratic regimes. Another factor is a long tradition of thinking, promulgated by Adam Smith and others in the Enlightenment, that market freedoms and commercial interests are the most productive and the least dangerous forms of social cohesion. As Jane Jacobs has argued, there are in human history two fundamental systems of survival, or ways of making a living, for human groups—taking and trading.[11] Economic survival through taking involves violence and conquest, and is associated with strong group ties and identifications, and with strong ontologies that justify the taking activities. Economic survival through trading historically involves engaging with strangers and members of other groups and relies on the development of less violent and more cooperative trust relationships. This is furthered by less intense belief systems and less stress on conformity with specific rituals, customs, and mores.

In a similar vein, economist Albert Hirschman surveyed early modern debates over the pros and cons of capitalism and found that many thinkers were aware of the difference between the socially destructive effects of what they called the "passions" compared with the constructive and salutatory effects of what they called "interests."[12] The concept of interest, so ubiquitous in our economic and political vocabularies today, was effectively transformed into an individualistic and consumptive idea. Intensely held spiritual or metaphysical beliefs and feelings would lead political life in self-destructive and wasteful directions. More calculating and material beliefs and goals would channel energies toward the productive exploitation of nature through trading and sharing knowledge and technology rather than wasteful coercion in an attempt to change the beliefs and behaviors of other cultures and societies.

The story proper of modern political thought, as well as modern economic thought, begins with the Reformation, and the terrible religious warfare of the fifteenth and sixteenth centuries. The great European intellectual reorientation known as the Enlightenment and secular liberalism as a political philosophy were born, at least in part, in reaction against this traumatic political experience. Historian Jonathan Israel summarizes the nature and significance of this reorientation as follows: "The Enlightenment revolutionized man's understanding of morality, history, law, politics, philosophy, science, and justice itself, by eradicating religious authority, theology and metaphysics from our evidence, reasoning and arguments—if not necessarily from our system of values and belief—and generating a searching, wide-ranging body of this-worldly social criticism."[13]

The scientific revolution and the Age of Reason kindled hope that humankind could move into a new age of peaceful coexistence. The medium of this was called "gentle commerce," which tapped into a deep wellspring of human desire for material security and comfort. If trading was the key to a progressive improvement in happiness and pleasure then, now in the Anthropocene the tables seem to be turned because our great economic machine of material production and consumption is in danger of turning on and consuming itself. The dangers and violence of taking as a system of survival in Jacob's sense are still with us, to be sure. But trading is becoming less and less a viable alternative, at least in its present dispensation. Is there no third way, no middle ground between a taking that is unjust (whether it be a taking from other societies or a taking from the planet) and a trading that is unsustainable? And if there is, how can we get there?

Ontology and Democracy

In regard to this second question especially, a paradox confronts us: History suggests that a "thin" theory of human flourishing and the weak materialistic and rationalistic ontology of market liberalism is all a liberal democracy can safely countenance. However, that may not be sufficient to motivate us to undertake self-imposed limits and constraints nor to accept the mitigations, adaptations, and even sacrifices that moving from a consumptive political economy to an ecological political economy will require. We need a Goldilocks ontology, one not so transcendental, strong, and thick that it will lead to a new Hobbesian state of nature among incommensurable worldviews, nor one so anthropocentric, skeptical, and thin that it lacks the power to motivate radical ecological conservation, one hopes under the auspices of a democratic governance of some kind.

In this middle ground there may again be two different postures that can be taken within the frame of an ecological social contract. Both may be able to thread the needle between strong and weak ontology and set up a symbiotic connection between enlightenment and enchantment, reason and imagination. Each is an available mode of re-enchanted democratic politics in an Anthropocene age. The first may be called "worldview democracy." It involves a reorientation of our culture and worldview—a transformation of our "spirit" as a political community in Montesquieu's sense, turning us from being a people of competitive consumption into a people of sustainable ecological responsibility. The second may be called "discursive democracy." It involves the institutionalization and empowerment of participatory and deliberative governance within a diverse and pluralistic society and culture, a system with multiple centers of order and coherence (i.e., a "polyarchy") and a system in which power and organizing resilience are widely diffused (i.e., a "panarchy").[14] This is the kind of democratic governance that grows directly out of action in concert with others, shaped by debate and deliberation in which strong ontological convictions and substantive conceptions of the good or contentments are not debarred from the public square.

There are some important similarities, as well as differences between these two types of democratic politics, and I shall return to a more detailed discussion of them in Chapter 12. For the moment, I turn to what can be called the danger of democratic despotism.

Democratic Despotism

Both worldview democracy and discursive democracy can be prone to a kind of democratic despotism that is as antithetical to freedom and justice as theocratic despotism, but more subtle. Writing in the early nineteenth century and in the aftermath of the French Revolution, and fascinated by the unique historical emergence of equality and democracy in the young United States, Alexis de Tocqueville was an acute student of the possibility of democratic despotism. He saw its roots in the deracinated form of individual consciousness and social life that was being brought about by an emphasis on equality, individual rights, and liberty, and the pursuit of happiness and interests individualistically defined:

> I think . . . that the kind of oppression with which democratic peoples are threatened will resemble nothing that has preceded it in the world. . . . I want to imagine with what new features despotism could be produced in the world: I see an innumerable crowd of like and equal men who revolve on themselves without repose, procuring the small and vulgar pleasures with which they fill their souls. Each of them, withdrawn and apart is like a stranger to the destiny of all the others; his children and his particular friends form the whole human species for him; as far [as] dwelling with his fellow citizens, he is beside them, but does not see them, he touches them and does not feel them; he exists only in himself and for himself alone, and if a family still remains for him, one can at least say that he no longer has a native country.

But Tocqueville feared that this very individualism would lead, not to the kind of negative liberty and space for idiosyncratic individuality that his contemporary, the English philosopher John Stuart Mill, envisioned, but rather to a new kind of paternalistic authoritarianism, now embraced by, not imposed on, privatized selves:

> Above these an immense tutelary power is elevated, which alone takes charge of assuring their enjoyments and watching over their fate. It is absolute, detailed,

regular, far-seeing, and mild. It would resemble paternal power if, like that, it had for its object to prepare men for manhood; but on the contrary, it seeks only to keep them fixed irrevocably in childhood; it likes citizens to enjoy themselves provided that they think only of enjoying themselves. It willingly works for their happiness; but it wants to be the unique agent and sole arbiter of that; it provides for their security, foresees and secures their needs, facilitates their pleasures, conducts their principal affairs, directs their industry, regulates their estates, divides their inheritances; can it not take away from them entirely the trouble of thinking and the pain of living?[15]

Tocqueville was a formidable political theorist and a great writer, albeit somewhat prone to hyperbole. Nonetheless his sobering notions have serious resonance even today. Referring to the "immense tutelary power" (by which he can only mean the state and the pressure of hegemonic ideas and values in the popular culture), he could as well have been talking about contemporary consumer culture: "it likes citizens to enjoy themselves provided that they think only of enjoying themselves."

If ecological democracy were to morph somehow into a kind of ecological democratic despotism, it would subvert precisely the civic and democratic promise of the ecological social contract. What Tocqueville sees is a democratic despotism that turns public citizens into private consumers of things to satisfy their own wants and of images and ideologies to reinforce their own self-identities. The contemporary philosopher Charles Taylor, commenting directly on Tocqueville's argument, writes:

The danger is . . . fragmentation, that is, a people less and less capable of forming a common purpose and carrying it out. Fragmentation arises when people come to see themselves more and more atomistically, as less and less bound to their fellow citizens in common projects and allegiances. . . . [F]ragmentation grows when people no longer identify with their political community, when their sense of corporate belonging is transferred else-where or atrophies altogether.[16]

Ecological Enchantment

Does secularization, understood as loss of belief in transcendent being, meaning, or purpose, inevitably carry with it such deracination? Is a sense of the sacred necessary to protect democracy—especially worldview democracy—from its own form of despotism? Taylor has perhaps done more than anyone recently to think through this complexity, as his magisterial work, *A Secular Age*, attests.[17] In an essay entitled "A Catholic Modernity?" Taylor explored both the tensions inherent in what I am calling enlightenment and enchantment, and their essential, although not always perceived, interconnections.

At one level Taylor is concerned with the question of how the Roman Catholic faith can come to grips with the challenge of secular modernity. Yet on another level, he is equally concerned with the question of whether the culture of modernity itself—with its rationalism, scientific worldview, technological power, individualism, tolerance, diversity, and separation of Church and State—can offer a new kind of "catholic" (universal) faith. In other words, can we find a "civic religion"—a faith of universal rights and values grounded only on humanism, without any form of transcendent theism whatever—at least as far as the public realm is concerned?

Taylor's answer is no. Modernity and secularization are insufficient: the faith of humanism is incomplete without something beyond the human to worship. But at the same time, humanism is essential. Without secularism and modernity, Taylor argues, religion alone would never, through its own unfolding, have come to affirm universal human rights.

> The very fact that freedom has been well served by a situation in which no view is in charge—that it has therefore gained from the relative weakening of Christianity and from the absence of any other strong, transcendental outlook—can be seen to accredit the view that human life is better off without transcendental vision altogether. . . . The strong sense that continually arises that there is something more, that human life aims beyond itself, is stamped as an illusion and judged to be a dangerous illusion because the peaceful coexistence of people in freedom has already been identified as the fruit of waning transcendental visions. . . . Do we really have to pay this price—a kind of spiritual lobotomy—to enjoy modern freedom?[18]

I would add that the notion that "there is something more, that human life aims beyond itself," is also a necessary aspect of the ecological imagination: seeing something larger than the human purpose and interests per se, placing the being and becoming of humanity in a larger context, is not exclusively a matter of transcendental or supernatural vision; it is also essential for an immanent or natural vision, properly understood.

Together with Francis Bacon and René Descartes, Hobbes was an important figure in the transition from a strongly hierarchical and authoritarian medieval social order and cosmology to the secular worldview. This paved the way, as Taylor notes, for the development of both human rights and market society and thus it may be said that secularism, if not logically necessary to, is nonetheless closely linked with, the consumptive social contract. Hobbes was born in 1588 as the Spanish Armada was approaching England, and as an adult he was horrified at the English Civil War that was bringing his own world down around his ears. Deep disagreements about religious orthodoxy and the requirements of salvation could be tamed under the banners of science and materialism. Questions of transcendent value and purpose must give way to questions about the limits of what human beings can know and the limits of their rational control of their own interactions and behavior. Hobbes was a master of modern suspicion and irony. He profoundly distrusted those he referred to as "ghostly men," who claimed to have knowledge of transcendent truths but were actually using this and the credulity of the faithful in order to consolidate clerical and political power. The true source of value and command was not divine; it was human beings acting to create sovereign authority and the collective commandment of the social contract. This was rooted in the natural within the human. The prime movers in human life are fear and disorder: "The passion to be reckoned on," Hobbes said, "is fear." We turn to religion not because we can know it is the truth (we can't), but because it promises security and peace. In the 1640s, it had not been making good on that promise. What is the solution? Consolidate political power, harness fear to productive, orderly ends, and the novus ordo seclorum will emerge.

To be sure, the secularization of the public realm did not catch on right away. Hobbes's *importance* was great, but his *influence* was limited and was overshadowed by Locke and the various thinkers of the Enlightenment in the eighteenth century. Revived by the utilitarians and economists in the nine-teenth century, Hobbesian thought, however, has lived on and has been appropriated into mainstream liberalism, domesti-cated, as it were, by milder thinkers in contemporary philoso-phy and political science.

Pitié and the Call of Suffering

Less than a century after Hobbes's death an equally powerful counterpoint to him had arisen in the work of Rousseau. Rousseau maintains that human beings need a sense of the transcendent and, to that extent at least, religious belief because they are fundamentally progressive, developmental beings with a built-in logic of moral and spiritual growth. (Unfortunately, this growth is a two-edged sword and can also spiral downward into degeneration.) They are not fearful beings, frozen in a pet-rified, universal Hobbesian present. And this sense of the tran-scendent is not merely a gullibility making them susceptible to manipulation and control.

For Rousseau the mind and will of human beings need a sense of something larger than themselves. Without this sense, with only a restless desire for power after power, human exis-tence will remain mean, nasty, and impoverished, if not alto-gether brutal and short, even in the sovereign state.

So Rousseau concluded that Hobbes's purely skeptical and secular solution would not work and that politics needs a measure of enchantment, a sensibility of meaning and value above and beyond our immediate present and our own per-sonal desires and vanities. If Hobbes held that the passion to be reckoned on was fear, Rousseau said that it was *pitié* (often translated in English as pity or compassion), which Rousseau defined as "an innate repugnance to see [a] fellow-creature suffer."[19] This is close to the ability to put oneself in the place of the other and to identify with the experience of the other. But Rousseau does not stress that the suffering in the other we pity is simply empathetic; he thinks of it as a deep repugnance within the observer that is more primal than imagined identification and is a visceral response to the presence of suffering in the world. Suffering calls us to respond; it is not a reasoned choice or discrete decision guided by an ethical principle or a calculus of collective happiness or individual self-interest. In this way Rousseau is talking about the ability of human beings to be motivated to purposive action by a capacity to think outside of themselves and by the recognition of a reality independent of themselves, which nevertheless significantly affects their own subjective awareness and experience. This is the connecting link between the human mind and self and a broader reality and context of meaning and value that I am calling the transcendent. It is also the connection between ourselves and that openness of thought and feeling—conviction and contentment—that I am calling enchantment.

If what he calls pity is a capacity to connect with larger being, Rousseau did not understand its workings in isolation but always in psychological interaction with the two other varieties of self-love, *amour de soi-même*, or the survival imperative, and *amour propre*, which takes us outside of ourselves into the illusory world of appearances and the unreliable and indecipherable social reactions of others, but not to a larger truth or telos. For Rousseau, *amour propre* was not simply a preoccupation with the self at the expense of others. It was, instead, a peculiarly other-directed preoccupation with the self, a drive to love not the self as such, but the image of the self as reflected in the mirror of others' attitudes, opinions, and responses. In the *Discourse on Inequality* Rousseau underscored these aspects of *amour propre* and made them the basis for very sharply differentiating between *amour propre* and *amour de soi-même*:

> *Amour propre* and *amour de soi-même* are two passions very different in their nature and their effects, and must not be confused. *Amour de soi-même* is a natural sentiment which inclines every animal to watch over its own preservation, and which, directed in man by reason and modified by pity, produces humanity and virtue. *Amour propre* is only a relative sentiment, artificial and born in society, which drives each individual to have a greater esteem for himself than for anyone else, inspires in men all the harm they do to one another, and is the true source of honor.[20]

In a parallel passage in *Émile*, Rousseau elaborated this distinction even further and made it the basis for the dual possibility of humankind's psychic development:

> *Amour de soi-même* is always good and always in conformity with order. . . . *Amour de soi*, which regards only ourselves, is contented when our true needs are satisfied. But *amour propre*, which makes comparisons, is never content and never could be, because this sentiment, preferring ourselves to others, also demands others to prefer us to themselves, which is impossible. This is how the gentle and affectionate passions are born of *amour de soi*, and how the hateful and irascible passions are born of *amour propre*. Thus what makes man essentially good is to have few needs and to compare himself little to others; what makes him essentially wicked is to have many needs and to depend very much

on opinion. On the basis of this principle, it is easy to see how all the passions of children and men can be directed to good or bad.[21]

In short, *amour propre* was the selfishness born out of social dependence and a self-identity symbolically constituted by the competitive comparisons inherent in an inegalitarian, status-oriented culture. It was the sign of a profound emptiness. Men of *amour propre* were hollow, constantly in doubt about their own identities, voraciously needing to be filled and vicariously living outside themselves through others. It was only through others that men of *amour propre* were able to achieve a "sentiment of their own existence."[22]

For its part, *pitié* does not require loss of self in a transcendent reality or a complete disregard of self-interest. It moderates and reshapes or redirects the reasoning logic of self-interest; in Rousseau's phrase it "tempers the ardor [a human being] has for his own well-being."[23] Hence in his quest to recover or carve out a space for enchantment, Rousseau did not intend to cancel out enlightenment. He sought, as a supplement to the social contract, an ethos or supportive underlying political and moral culture that would animate it with a sense of larger being. This was a civic faith, ultimately a democratic faith, not a religious theology.

As a result of his sociological and psychological views, Rousseau felt that a cultural supplement of enchantment was needed to make the rational enlightenment of his notion of a democratic social contract work. But that was not the only reason the dimension of enchantment is necessary for him. The internal logic of his conception of the social contract and sovereignty itself depends in a fundamental way on the capacity to recognize and connect with transcendence. This has to do with his concept of the general will. Rousseau was a political theorist who took democracy seriously—one of the few, in fact—but he was not a majoritarian democrat. He explicitly differentiated the general will (*volunté générale*) from what he called "the will of all" (*volunté de tous*). The general will is a normative concept, not a numerical or procedural one. It has to do with the spirit, the intention, the moral orientation of the virtual collective will of a political community. Kant later interpreted the general will as impartiality and made it a category of the possibility of reason itself. Rousseau said a number of things that encourage such a reading, but I think it is more fruitful, at least for our purposes here, to see the general will as a gesture in the direction of relational being and an ontological condition of symbiosis and interdependence. Our potential capability to will or intend the flourishing of others with equity and impartiality is inseparable from the possibility of flourishing ourselves.

The kind of political order within which human beings can fulfill their moral and spiritual potential—which Rousseau called a "republic"—requires a sense of community and a deliberate, reasoned conviction to serve and sustain the common good that one shares with others in the political community. Like Hobbes before him, Rousseau believed that there needs to be a consolidated power or authority to hold the political community together. But for Rousseau this consolidated power, this Sovereign, does not rule on the basis of fear or calculated self-interest; it rules by moral conviction and conscience. And for Rousseau the Sovereign is not a person or even a body of persons: it is not a king, a parliament, or a democratic assembly as such. It is the moral spirit, the driving force of a political culture and a way of life.

Much of what keeps this commitment to the common good or general will alive is the activity of republican citizenship itself—active, interested, engaged, caring. But Rousseau (who was not an optimistic thinker) did not think that the praxis of citizenship alone would be enough. In the first place, he felt that it would be necessary to sustain it with a domestic sphere based on the influence of a strong matriarch, a "republican mother," whose family-centered role left her no time and no place in the male-dominated public sphere, but whose cultural function was deeply and fundamentally political.

Moreover, he felt that it would have to be sustained by the institution of a civic religion to which all citizens (both *citoyens* and *citoyennes*) would have to profess their belief, regardless of what their other, private religious beliefs and practices might be. Rousseau nowhere assumes, much less argues, that private religion will wither away. And he was just as keen as Hobbes had been to keep private ontological belief in check insofar as political behavior was concerned, but for rather different reasons.

Rousseau's understanding of civic enchantment offered people what they yearned for in terms of transcendent meaning and spiritual fulfillment, but it did so under the terms and conditions of what we would now call "human rights," especially equality of civic respect and toleration. Where does this civic enchantment come from? Not from visionaries or a priestly class. It is the distillate of a culture and a tradition; it comes from the moral learning of enduring ways of life, molded like a dynamic ecosystem across time and generations in response to changing landscapes and challenges. Rousseau's civic religion, what he referred to as *moeurs* (morals and manners) or laws written in the heart, is in constant dialogue with human becoming.

Rousseau tropes this as a personification; the source of civic religion is the figure he calls the "Lawgiver" (*le Législateur*).

He has in mind the ancient examples of Moses, Lycurgus, and Numa. Rousseau's Lawgiver does not found the political system, he founds the culture, the *moeurs*, upon which the ideal political system will later depend for its proper functioning and sustainability. The Lawgiver brings the implicit values upon which a community of human self-realization can be founded. A Rousseauian civic religion does not codify those values (Rousseau does not have his Lawgiver actually give the law or bring back tablets written by God) so much as it provides a form of practice that can embody those values in the lives of the citizenry. Having no power in his own right, the Lawgiver forms the shape that the use of political power will take. Unlike his ancient models, Rousseau did not adequately develop the implications of his ideas about the general will or civic religion, nor did he provide institutional details to capture and contain his political vision.

Earlier I suggested that a democratic solution to the dialectic of enlightenment and enchantment was either a worldview democracy or a discursive democracy. As I interpret him, Rousseau resists those alternatives. He in fact anticipates what we might think of as a worldview culture coupled with a discursive democratic politics. This is a recipe for the re-enchantment of an ecological social contract. He himself calls it a civic (or civil) religion, which offers a transcendent orientation without being otherworldly or supernatural. It is a religion of the commons and the land (in Leopold's sense) and the rule of law. It is a substantive conception of right relationship and right recognition, not simply an expedient or prudential one designed to keep the peace. Precisely because religious belief and faith are authentically important to people, and because religious sentiment can readily become intolerant and authoritarian, it is all the more important that an ecological social contract couple strong spiritual contentment with a strong philosophical commitment to impartial justice, public deliberation informed by scientific knowledge, and a sense of human membership and plain citizenship in the surrounding biotic community. Rousseau's notion of a flourishing republic reminds us that if we cannot become fully human in the absence of a right relation to the political community and a just recognition of our fellow humans, then neither can we do so in the absence of a caring, communing relationship with nature around us.

Part III

Terms of an Ecological Contract: Humans in Nature

It is scarcely necessary to remark that a stationary condition of capital and population implies no stationary state of human improvement. There would be as much scope as ever for all kinds of mental culture, and moral and social progress; as much room for improving the Art of Living, and much more likelihood of its being improved, when minds ceased to be engrossed by the art of getting on.... Hitherto it is questionable if all the mechanical inventions yet made have lightened the day's toil of any human being. They have enabled a greater population to live the same life of drudgery and imprisonment, and an increased number of manufacturers and others to make fortunes.

—John Stuart Mill[1]

[P]rogress, as we have come to understand it [in the modern age], means growth, the relentless process of more and more, of bigger and bigger. The bigger a country becomes in terms of population, of objects, and of possessions, the greater will be the need for administration and with it the anonymous power of the administrators.... Bigness is afflicted with vulnerability; cracks in the power structure of all but the small countries are opening and widening. And while no one can say with assurance where and when the breaking point has been reached, we can observe, almost measure, how strength and resiliency are insidiously destroyed, leaking, as it were, drop by drop from our institutions.

—Hannah Arendt[2]

Man has survived hitherto because he was too ignorant to know how to realize his wishes. Now that he can realize them, he must either change them or perish.

—William Carlos Williams[3]

6

Agency, Rules, and Relationships in an Ecological Social Contract

Thus far I have introduced the distinction between a social contract of consumption and an ecological social contract, but the discussion has mainly focused on the more general form and logic of social contract thinking. A social contract, regardless of its specific normative content, has important social and cultural functions and provides a frame within which people work out the conditions of their convictions and their contentment—their values and identities, their sense of right relation (justice and care) and right recognition (voice and respect) with other humans and nonhuman forms of life. The social contract is also a framework within which people form their sense of a good life for themselves and their community. The good life is a life of developmental attainment and flourishing, a life of dignity, happiness, gratification, and fulfillment. When under specific historical, cultural, and environmental circumstances, a social contract frame is invested with substantive conceptual and ethical content, then all these terms I have just listed pertaining to the right and the good take on a particular interpretation and then become more directly linked to specific political, economic, institutional, and power arrangements. For example, social contracts with substantive normative content are connected with political economies and governance structures.

Spelling out the substantive content of the ecological social contract with appropriate research in moral and political philosophy and social science is not the work of a single book, by any means, and goes beyond the scope of what I offer here. But in Part III I identify some of the "terms" of an ecological social contract, that is, some significant facets of what a fully realized substantive ecological social contract might look like. The facets themselves are not particularly original. But I shall introduce some conceptual distinctions and ideals that are not commonplace in most discussions of these topics in an effort to stimulate thinking about political and economic activities in new ways. And my aim is not to complete the conversation on new ways of understanding these facets but to contribute my own voice and sense of moral direction to an already broad and lively ongoing discourse.

This chapter presents substantive notions concerning intentional agency, life experience, and relational social practices. Its purpose is to lay the groundwork for the following chapters that take up more specific ideas in the economic and political domains, in which I discuss new ways of thinking about economic wealth, property and our relations to what we use, freedom, and citizenship.

Redrawing the Moral Map of the Ecological Contract

An ecological social contract is less about how we do act and interact with other people and nature than about how we should act. But where does this "should" come from? In part it comes from what we objectively have to do in order to survive, given what science tells us about the conditions of survival, and in order to protect those very qualities that are most distinctive in our humanness. However, the normative "should" of the ecological social contract is not all negative and precautionary, it is also affirmative. It comes from what we have the potential to do in order to make our humanness more profound and to extend the capabilities necessary for flourishing to all persons.[1] These capabilities are to be used in ways that are sustainable and compatible with the species-appropriate flourishing of other forms and systems of life.

So at the center of an ecological social contract are ethical principles, rules, and norms, which, when they are accepted and practiced widely enough, become an orienting and directing force in a society. Now, such rules certainly do grow out of historical traditions and already prevailing systems of morality. But an overly conservative inference should not be drawn from this. The orientation offered by the substantive norms of a social contract is not solely tacit or habitual. Persons behaving intentionally and reflectively—that is, persons exercising "agency"—also make these rules and reflect on their broader meaning and justification. Moreover, when historical shifts occur in social contracts, the resulting transformations and reconstructions of outlook and values can be radical and systematic.

As I write this I am looking at a reproduction of a sketch by Leonardo da Vinci called the "Vitruvian Man" (*Le proporzioni del corpo umano secondo Vitruvio*) dating from around 1490. It is a familiar image and probably would elicit broad and immediate recognition. For me it is a kind of universal icon, akin to the NASA photographs "Blue Earth" (showing a floating, beautiful, cocoon of life) and "Dark Earth" (showing the distribution and concentration of power on the planet at night).

The Vitruvian is the image of a male human form with arms upstretched and legs spread wide apart so that the body forms an X within a circle. (A circle is the traditional figure of perfection in Western culture; it is God's shape, and according to Origen, the shape of the soul.) This figure's feet are firmly planted on the ground, and yet the upward gestures of arms, head, and overall posture—man the biped—give it the aspect of a creature of vision and longing. And this is not its only duality. I said it looks like the letter X in the Roman alphabet, this body stretched out within a wheel, but it also resembles the Greek letter *chi*, which signifies the meeting point of the two opposing lines of human life—the intersection of happiness and misfortune, good and evil, life and death.

This is an appropriate image to use as a starting point for moral mapping. The human person has four basic points and 360 degrees on the ontological circle that surrounds her and demarcates her moral world. Up, down, right, and left. These points are on two axes or planes, the vertical and the horizontal. Now imagine the person within not a flat circle, but a three-dimensional one, a sphere. And imagine that the person is not locked in one posture, with the head up and the feet down, for example, but unhinged and free to tumble and swivel around in the sphere.

Now we begin to approach something like the existential and ontological situation of the human moral life framed by the ecological social contract. Morality operates on both the vertical and the horizontal planes of our experience and our being—it is anchored in relation to the earth like Antaeus, who is invincible as long as his feet are touching the soil from which he springs, and morality gazes at the heavens above like Daedalus, who invented (among other things) wings enabling him to fly, and who was wise enough (unlike his son, Icarus) not to use them to fly too high. It also orients, as I have said, the self horizontally in relation to other human persons. As we tumble, sometimes out of control, we may kick others with a foot rather than offer them a helping hand, and it may not always be our head that we lift up to heaven.

But tumble we must. And we must learn to get control of ourselves so that we can move around purposefully in the sphere of our moral life. Above all, the moral life is a scene of human action and individual human agency. The moral life is a place of doing; more precisely, it is a place of *being as doing*. The concepts of "agency" and "action" refer to doings that are intentional, meaningful, and purposive. Action is a communicative and expressive performance that is comprehensible within a cultural framework of shared meanings. Action is controlled, deliberate flying, not tumbling head over heels. Action is not the same as "behavior" or bodily movements that can be explained without reference to intention and the person's voluntary control. Behavior that is solely the result of a reflex, a conditioned response, or is wholly determined or coerced by forces outside the agent's control is not action in a morally relevant sense of the term. A wink is an action; a blink is not.

I don't mean to suggest that individuals are agents, purposive, deliberate, in control, all of the time. There are certainly times in human life when this kind of talk makes about as much sense as it does to talk about piloting a roller coaster, or herding cats. We go where life takes us; we don't take it.

To see the ecological social contract in terms of moral rules and promises to keep is not the whole story. It does contain promises to keep, but also promise yet to be discovered. Morality is a burden, a weight primarily when it is seen as a structure of rules, commandments, obligations, and requirements. But there is another way of looking at morality that permits it to permeate the whole of one's active life without these oppressive consequences. The moral core of an ecological social contract can be regarded as a way of living one's life and a way of being in the world that one can come to take on as lightly and unoppressively as a smile or a style. The convictions and contentments of an ecological social contract and an ecological way of life are not limited to hard choices or self-sacrifices only; they can be woven into the fabric of our lives or placed into the current of its ebb and flow.

In short, the ethical substance of the ecological social contract is in part a creed and a set of commandments that provide rules and directions for living, but only in part. Morality is not just obedience, it is also exploration. Much of lived moral experience consists of the ongoing task of finding one's way, of taking one's bearings, and of locating oneself in relationship to others, in relationship to our own conscience, and in relationship to purposes and levels of being wider than our own. Morality depends on finding the proper orientation, on knowing who and where one is, and in discerning the right direction in which to travel. It is not merely a response to pre-established rules, it is an active engagement with those rules, an active membership in an ongoing community of interpretation of those rules, their application and meaning. The moral life is a drama of both discovery and creation, accommodation and innovation, receiving and reshaping, or revisioning. A moral education is learning to read and follow one's compass around the geography of being and becoming human in nature. It is learning not merely by reading the warning signs or rules that others have posted, but also by leaving the well-marked trails and coming to know its landscape, its terrain, its texture and ecology.

The moral life is a journey of discernible destination, but unknown route. Most of the time it takes us down well-worn paths, but there are moments in every moral life when the path seems to suddenly disappear in a thicket, or when we simply have the curious urge to go off another way. When exploring unfamiliar domains, some basic equipment is needed, and once again we have to resort to metaphors to convey the qualities and capacities for moral life that are constitutive of, and inherent in, the human being as person, self, and agent. In dangerous terrain, one needs a sense of orientation and direction, a sense of danger and caution that keeps one from running into pitfalls and from going over the edge of cliffs. One develops a sense of what needs to be paid attention to and what does not, and an ability to recognize when you have reached your destination and have come home. These are the four R's of the moral core of an ecological social contract: reckoning, restraint, responsibility, and respect.

Reckoning or a Moral Compass

Moral reckoning provides a sense of location in relation to principles or rules of transcendent being and universal lawlike (nomothetic) values, and also in relation to rules of being and becoming in place and particular, contextualized (ideographic) values. This sense of self-location is the first moment of the moral life and the moral point of view. Ethics is corrected vision; it is a proper awareness of one's location both outside and inside time and place. Acting unethically is so often a function of being disoriented or lost; of forgetting or losing track of where we are and what is appropriate given where we are. We don't have anything to look up to because in this state of vertigo we have lost our orientation, and we don't know which way is "up" anymore.

Restraint or Moral Brakes

The moral life is anchored in the sense of going too far, a sense of trespass. All human cultures engage in some form, however varied, of fighting or aggressive behavior as a part of their social lives, but they all react viscerally to the drawing of first blood. A dangerous line has been crossed. Morality is born at that red line. Human beings are free-range creatures by nature; we are curious, even avaricious in our desire to explore, to grasp, to appropriate. *Pleonexia*, the Greek word for taking what is not your own, for grasping and violating proper boundaries, is the prototype of all immoral conduct. Without moral brakes, we run amok. With them, we can respect boundaries. Without this pillar firmly in the mind and in the heart, rules may still be devised, but they may be self-defeating. Without the moral brakes of restraint, moral rules too readily come to be seen as mere hurdles, mere challenges to the clever to circumvent, and not, as they should be seen, as warning signs and fences in front of a yawning abyss. Earlier I spoke of taking and trading as two prototypical systems of survival. Moral brakes and boundaries don't pertain only to taking, they are essential to trading as well. The social contract of consumption has lost touch with its necessary limits.

Responsibility or Moral Solidarity

By moral responsibility I mean not so much duty, obligation, or blame, but rather the *ability to respond* and to establish a connection between oneself and some Other(s). I believe that our moral sense and all our ethics are ultimately rooted in this capacity to have our own deepest motivations and sense of ourselves moved by the recognition of the Other. Or, to put the point more precisely, the moral life is rooted in the recognition of the mutual "being-there" that one shares with the Other (both human and nonhuman) as a living being: namely, the condition of mortality, vulnerability, and sheer insufficiency. This alone saves us from solipsism and narcissism, which are the negations of the moral life. Moreover, in this way, we learn to embrace the forms of difference to which we cannot properly be indifferent. Simone Weil refers to this reaching out as the capacity to be attentive. To attend: this suggests patience, listening, service. "The capacity to give one's attention to a sufferer," Weil observes, "is a very rare and difficult thing; it is almost a miracle; it *is* a miracle."[2] Attention not only properly orients our vision, it expands it. Morality has but a limited triumph over narcissism when it leads us to have regard for those most like us or those who have directly helped us. It is the Other as Stranger to whom we must bind ourselves; it is the Stranger who calls our name. The moral life consists largely in learning the Other's name. By encountering the transcendent, we are assisted in our calling to name and to be named by other people, and to be there, with and for them in the horizontal moral community.

Respect or Moral Community

Community is a pillar of the moral life for it provides the place, or grounding, and the time to fulfill the promise of being there with and for others. The moral life is all about how well we can see, hear, and feel these things. We on the horizontal landscape of human-to-human or humans in society relations must establish moral community on this dimension. But what about the other dimensions of our relationality? Respect and community allow us to come out of ourselves and go out to one another. By going out, we can also go "down" to develop a moral relationship of right recognition and respect for nonhuman Others and the natural world. By going down, we also go up.

7

Wealth

From Affluence to Plenitude

The twenty-first century thus far has been a time of war and economic crisis. It has also been one of ideological polarization and political paralysis. In the United States, the Affordable Care Act, a historically significant achievement of governance—health care reform intended to increase access and social justice—has not been cause for collective celebration and affirmation to say the least. A globally significant challenge—slowing and reversing the deadly process of climate change—has not united us in common effort and common purpose. A pervasive atmosphere of distrust has arisen from which even scientists are not immune. Business as usual, politics as usual, and even reason as usual seem to have become derailed and unhinged. Ways of coping with instability that have served our society in the past—such as government fiscal stimulus, regulatory reform, international negotiation and coordination, appeals to personal conscience, philanthropy, and responsibility—seem unavailing. It is difficult to avoid the suspicion that we are losing our grip on something fundamental, yet it is also difficult to say what—our optimism? Our resilience? What have we wrought? What kind of world are we squeezing into (or out of) shape?

Incremental reforms and tinkering with the plumbing of our political economy will not suffice, yet that is all our leaders and institutions are capable of achieving at the moment, if even that. Let's talk about something different and something more. Let's talk about imagining an economy otherwise, based on a new story about ourselves, our duties, and our destinies. We surely are creatures of need and desire, agents of exchange and competition, pursuers of self-interest and givers of mutual assistance. Yet, as economic beings we can be a far cry from the *homo economicus* of the liberal market mentality, global neo-liberal capitalism, and the consumptive social contract. Let's explore some facets of what it might be like if we were to act like different kinds of economic being. To get this discussion started, I propose in this chapter to reflect on the ideas of scarcity and its flip side, affluence; plenitude and its flip side, frugality, and the notion of going without ecologically unsustainable economic growth.

Musical Chairs

When it comes to rethinking assumptions and values, there is no better or more fundamental place to begin than with the ideas of scarcity, affluence, and economic growth. Today we live with an ever-present awareness of scarcity. There are not enough jobs. There is not enough money to keep pace with rising prices and taxes. Necessary services are often not available. We are running out of credit. We are running out of oil and the time needed to develop sustainable sources of energy. We are running out ... We are running ...

This scarcity is paradoxical. Scarcity sits side by side with material affluence. Indeed, our great stock of possessions may even reinforce our sense of scarcity. The key to this paradox is the realization that scarcity and affluence are not only economic issues, they are also moral and psychological ones. They are not primarily about how much we have, but about how we relate to others and how we live. Scarcity and affluence are two sides of the same coin. They preside over a culture of competitive advantage; a culture of beggar thy neighbor. Children are taught the lessons of scarcity and affluence at an early stage, as with the kindergarten game "Musical Chairs." In this game chairs for each child are arranged in a circle, and while the players are dancing around them in a circle, one chair is taken away. When the music stops the children must scramble to get a seat, which have now become scarce. Someone—the slow, the polite, the gentle—will be left out. Our economy today is Musical Chairs writ large and played on a global scale.

Affluence versus Plenitude

Scarcity and affluence should be distinguished from their ecological opposite numbers, frugality and plenitude. Frugality can bring people together in closer connection. Frugality, as in the times of rationing and hardship during World War II, actually brought about a sense of solidarity and connectedness; everyone had to help one another because they were all in the same boat. Plenitude can have similar community- and solidarity-building effects. This is because frugality and plenitude are experienced in common from the perspective of the first-person plural. Scarcity and abundance are experienced from the perspective of the first-person singular. Scarcity drives people apart and makes us anxious and insecure, and so does affluence. Neither scarcity nor affluence today are conditions of society or spirit that we genuinely share; they are merely ways of living that affect (or afflict) all of us. No one likes scarcity, but the recognition is dawning that affluence too is a way of living that is humanly unsatisfying and naturally unsustainable.

The word "plenitude," Juliet Schor observes, "calls attention to the inherent bounty of nature that we need to recover." Moreover, she argues, "It directs us to the chance to be rich in the things that matter to us most, and the wealth that is available in our relations with one another. . . . It puts ecological and social functioning at its core, but it is not a paradigm of sacrifice. To the contrary, it involves a way of life that will yield more well-being than sticking to business as usual, which has led both the natural and economic environments into decline."[1]

The great religious traditions of the world speak of another economy, an economy of plenitude and abundant being. And the contemporary science of thermodynamics provides support to the idea that life on earth itself is plenitude, for through processes such as photosynthesis living things increase energetic order.[2] An economy of plenitude is a counter-entropic economy where chairs are added, not taken away; it is an economy of life, justice, and solidarity as an alternative to the economy of appropriation. Precisely because we are so enmeshed in the economy of scarcity/affluence and getting, it is very challenging and complex for us to see the economy of frugality/plenitude and giving.

Above all, it is difficult for us to focus on how authentically we connect rather than on how successfully we acquire and consume. Another economy of frugality/plenitude is not primarily about possession: having or not having; it is about relationship: caring and sharing. Plenitude is not about how many things we have; it makes relationally meaningful what we create or earn. Frugality is not about how few things we have but about how carefully they are used.

The political economy of the ecological social contract is where making a living in an earth household becomes not an activity that sets us competitively at odds, but one that expresses our fundamental natures as ecological beings, interdependent in the face of our mortality, our finitude, our vulnerability, and our insufficiency. Sharing makes small portions filling. Selfishness, grasping, makes large portions unsatisfying. The plenitude that does not diminish the more it is shared is what truly overcomes scarcity in a way that affluence through economic growth never can. The plenitude that makes possible a community of compassion and caring even in the worldly condition of frugality is the exact opposite of our society's extended game of Musical Chairs.

Unfortunately, we believe—because we have been relentlessly taught—that the neoliberal economy is real and the other economy is merely a dream. At some point very large numbers of people in democratic societies are going to have to turn that upside down and say that frugality/plenitude are the givens, the natural realities of the human condition, while it is scarcity/affluence that are historically contingent nightmares from which we are trying to awake.

Reinventing Growth

Yves-Marie Abraham has written that "degrowth is a call for a radical break from traditional growth-based models of society, no matter if these models are 'left' or 'right,' to invent new ways of living together in a true democracy, respectful of the values of equality and freedom, based on sharing and cooperation, and with sufficiently moderate consumption so as to be sustainable."[3]

The concept of degrowth (*le décroissance*; *decrecimiento*; *decrescita*) is currently being used in a way that is deliberately imprecise.[4] I take it to be related to, but distinct from, economic ideals such as steady-state economics, social ideas such as decentralization and localization, and cultural ideas such as the contemporary agrarian movement. I interpret it not to mean no growth, but a different growth—a turning away from certain ways of defining and measuring "economic growth" and away from certain conceptions of happiness and fulfill-ment, which were in essence reinvented, for the worse, with the rise of capitalism and the discipline of economics in the eigh-teenth century. A different growth is not incompatible with full human employment, nor with human use of the natural world to meet human needs and to support the development of human capabilities and functionings. It will use nature in ways that accommodate the integrity of natural systems and in ways directed by ecological and thermodynamic under-standing of the limits, costs, and benefits of our activity. Eco-nomic incentives will be realigned in those terms. As currently defined, economic growth is an increase in the gross domestic product (GDP) of nations, which is a measure of the market value of all goods and services produced in a given time period. It is well known that this metric includes and excludes the wrong things and orients policy in inappropriate direc-tions. It is seriously out of touch with actual human welfare and happiness.[5]

In short, the notion of degrowth connotes a particular normative vision of an entire society. That said, the question arises concerning the kind of political economy and gover-nance that would be most fitting and best suited to a degrowth society. The reality of ecological limits and planetary bound-aries to major forms of human economic and technological activity—especially to those actions that are conventionally counted as economic growth—poses a normative and practical challenge to governance on national, regional, and global levels. We must countenance the possibility that liberal democracy, as we know it, will not be able to meet that challenge and so must give way to a new structure of governance.

It remains to be seen whether this transition to a post-growth governance will be done incrementally and in an orderly way, or chaotically in response to significant ecological crisis. It also remains to be seen what general form new regimes of governance can take—how representative and accountable governing officials and bodies will be; how limited their power and authority will be by constitutional and institutional mechanisms and by norms regarding due process of law, justice, and human rights; how democratic they will be and in what sense of the term.

8

Property

From Commodity to Commons

In Rousseau's view, the developmental pathway leading to human political and social being had to lead through the invention of property, which manifested itself first in the enclosure of land:

> The first person who, having fenced off a plot of ground, took it into his head to say *this is mine* and found people simple enough to believe him, was the true founder of civil society. What crimes, wars, murders, what miseries and horrors would the human Race have been spared by someone who, uprooting the stakes or filling in the ditch, had shouted to his fellows: Beware of listening to this impostor; you are lost if you forget that the fruits belong to all and the Earth to no one![1]

A central term of the ecological social contract involves rethinking the concept of property from the point of view of the way in which it locates human beings in relation to the material world and the effects it has on our moral imagination and conduct. Aldo Leopold, for instance, maintained that "[w]e abuse land because we regard it as a commodity belonging to us. When we see land as a community to which we belong, we may begin to use it with love and respect."[2] This chapter is a reflection on this fundamental insight.

The concept of property is fundamental to an understanding of the relationship between humans and nature. Property is not a thing; it is a relationship with objects and people, and hence it has consequences for others; it affects individual and group motivation and action; and it is value-laden, not value-neutral, from both an economic and an ethical point of view. Legal scholar Jeremy Waldron has noted that property involves "a system of rules governing access to and control of material resources."[3] In the Western tradition at any rate, going back to ancient Roman law, property is linked to the concept of rights. In modern times one privatized and individuated understanding of property links it closely with commoditization and market exchange, but that conception of property is not the only possible one. Most generally understood, property concerns access to resources, differentiating those who have free access to something from those who do not, and setting the conditions under which various individuals and groups may obtain access and a right to use. Often with the right of access and use come corresponding duties and obligations.

It is important to distinguish between private property and collective or common property. Today the term "property" is often taken to be synonymous with private property or individual ownership, but this closes off creative possibilities, especially in connection with sustainable land use and ecological trusteeship. Private property puts one person in control of how a resource is used; common property involves shared control and shared use. Indeed, there are forms of property rights in which the private owner does not have complete and exclusive control over access and use of a resource. Usufruct (*usus et fructus*, "use and enjoyment of fruits") arrangements cover a situation in which individuals have rights of access to property owned by someone else, as long as the property is maintained appropriately. Use and enjoyment rights to someone else's property historically have come in many forms and varieties, but one important notion that was developed over time is the idea of estover (*est opus*, "it is necessary") rights under which owners could not deny nonowning occupiers access to resources needed to sustain themselves and to perform their services on the land. Such resources could include access to grazing land, firewood, wild fruits, game, and the like. Note that common property involves shared ownership and shared power to determine resource use, and thus the normative dimensions of participatory decision making are readily apparent, but even private property ownership can be constrained and limited by normative notions—such as appropriate maintenance and usage—that are necessary to sustain people or ecosystems.

When property rights and relationships take the form of a commons, it is useful to distinguish arrangements of ownership from arrangements of access and use. Common ownership is compatible with very limited or exclusive use, while, as we have seen, private ownership is also compatible with a common dimension of more or less open access and use. Common-pool resources are those for which open access is difficult to restrain, either for physical or traditional cultural reasons. Thinkers such as Garrett Hardin and Elinor Ostrom have focused attention on the vulnerability of common-pool resources to overexploitation or neglect. This is a situation in which if the individuals involved follow the logic of rational self-interest, it will lead to suboptimal collective results.

Many, including Hardin himself, have drawn the lesson that privatization of the common resource is the best solution to this collective action problem. The economist Ludwig von Mises sees this problem as arising from commoners' disregard of external costs: "[If] the land is not owned by anybody," he writes, "it is used without any regard to the disadvantages

resulting . . . [and] other impairments of the future utiliza-
tion are external costs not entering into [the common users']
calculation of input and output."[4] However, Ostrom does not
join in embracing privatization as a solution to the degradation of the commons. Instead she sees in the localized, culturally informed participatory management of common-pool
resources a proven form of sustainable governance that avoids
the conventional approaches of competitive market privatization, on the one hand, and central government regulatory
and legal control, on the other.

Remaking the Old World

In his classic work *The Great Transformation*, historian Karl
Polanyi traced the changes that led in the late medieval and
early modern period to viewing land, human labor, and capital as commodities that could be bought and sold in an impersonal market. He regarded this way of looking at land and
labor as artificial and pernicious, but recognized how historically and politically powerful this alteration of perception had
been in history. "The commodity fiction," he wrote, "supplies
a vital organizing principle in regard to the whole of society
affecting almost all its institutions in the most varied way,
namely, the principle according to which no arrangement or
behavior should be allowed to exist that might prevent the
actual functioning of the market mechanism on the lines of
the commodity fiction."[5]

This had two fundamental consequences for the political
economy at the time. First, it changed the ways in which the
relationship between human beings and the material world
was understood and the ethical rules governing it. It also

changed the understanding of the relationship between an individual and his or her own body and mind, which were the sources of the labor that now could be exchanged in the same way. Finally, it created an entirely new and expanded system of money lending, thereby making the newly mobile wealth based in alienable land and labor more efficient to increase and manipulate. Investments in new technologies and commercial activity of many kinds were made possible by this change in the perception of what was right and permissible under the rules of society.

The second fundamental consequence that Polanyi sees in this development was that it fractured the way that economic production and consumption had been embedded in a larger cultural structure of meaning and norms.[6] This set the political economy apart as a semiautonomous sphere of life and activity, with rules and a logic of its own, and potentially in tension with the values, functions, and rules of other parts of the cultural system.[7] Polanyi argued that this commodification of material life and separation of economic activity from a more seamless cultural web of meanings, despite its material benefits, was in other ways impoverishing and diminishing to humanity. At the beginning of the twenty-first century, it increasingly looks like he was right. Writing shortly before his death in 1884, Karl Marx, for one, hoped and expected that by this time the future situation would be quite different:

> From the standpoint of a higher socio-economic formation, the private property of particular individuals in the earth will appear just as absurd as the private property of one man in other men. Even an entire society, a nation, or all simultaneously existing societies taken together, are not the owners of the earth. They are simply its

possessors, its beneficiaries, and have to bequeath it in an improved state to succeeding generations as *boni patres familias*.[8]

The relationship between humans and the natural world in principle has many dimensions and facets. The land has many vital functions. Commodification in a separate sphere of market exchange and merely instrumental economic use flattens the meaning of nature and perhaps removes some of the inhibitions against its inappropriate and ultimately self-defeating exploitation. Polanyi expresses the point this way:

> The economic function is but one of many vital functions of land. It [land] invests man's life with stability; it is the site of his habitation; it is a condition of his physical safety; it is the landscape and the seasons. We might as well imagine his being born without hands and feet as carrying on his life without land. And yet to separate land from man and organize society in such a way as to satisfy the requirements of a real-estate market was a vital part of the utopian concept of a market economy.[9]

The great transformation of which Polanyi speaks ran its course between the fifteenth and the nineteenth century in Europe and was extended through Western colonialism, which was the first wave of globalization. It not only made the economy a quasi-autonomous action system, loosened from many traditional cultural and religious normative constraints, it also ushered in a new cosmology or world picture.[10] New conceptions of the individual or self, new norms and expectations, and new forms of knowledge and technology all transformed the human perception of the natural world.

The social contract of consumption emerged out of this long process of modernity and modernization—the slow working out of the Renaissance, the Reformation, the Enlightenment, market capitalism, and globalization. The ontological separation of human life and well-being from natural living systems on local, regional, and planetary scales is now the moral default setting. It is the presumption that must be rebutted. When one is pleading the case for the planet, any effort to bring about closer coordination among these spheres, or to interfere politically or legally to reintegrate the economy with other cultural and natural systems, must bear the burden of proof against this presumption.

Indeed, the very concept of a public domain interconnecting these spheres, the unity to their plurality—or any notion of a commons or common domain that might be superintended by public intentions and goals rather than private ones—is jeopardized. In such a world, ecological trusteeship through democratic governance is not a self-evident truth, and it requires hard work to make a case for its ethical justification that garners the necessary popular support. But nature is chiming in and pressing its case, and can't be patronized much longer.

Today this state of separate action spheres is such a commonplace lens through which we understand the social world that it comes as a surprise to many that they have not always existed. When Leopold, writing in the 1940s, points out that we regard land as a commodity and as private property, it is no mere happenstance perspective that he is calling into question. Leopold employs a rhetorical style of writing that smooths out the flow of change so that the transition from the commodity and ownership construct to the alternative notion of membership in a biotic community is presented, not as a precarious possibility, but as a natural next step in a progression of human self-understanding and morality. He does not say "*If we were to see* land as a community to which we belong . . . ," but rather "*When we see* land . . ."

I assume that he was aware of the history that Polanyi (who was writing at about the same time) traces, at least in a general way, and I expect that he did appreciate what a struggle it would be to depart from that historical trajectory and the grip of market ideology. Perhaps he was caught up in the American progressive temperament. Perhaps he allowed style to take precedence over precision. Or, more intriguingly, perhaps he did not comprehend this as a linear sequence at all. Perhaps he thought of it as a coexisting pair of social constructions of reality, like two perspectives in those ambiguous figure drawings in Gestalt psychology—a rabbit or a duck; a face or a vase; an old woman with a scarf or a young woman in a hat—that a person may be able to shift back and forth between through an act of mental concentration and will. At least, most people can do so once they have had those two different pictures (each arising from the same lines, angles, and shadings) pointed out to them. Whatever Leopold intended, this is what I am trying to do with my suggested distinctions and dichotomies in this work.

The Idea of a Commodity

The essence of the way property figures in terms of the ecological social contract has to do with the question of how property confers the right of human beings to exercise what political theorist C. B. Macpherson has called "extractive power" within the natural and the social worlds, and the limits on that right. Is there anything that money cannot buy or anything that cannot be "owned"?[11] I don't mean just practically—technological limits prevent one from owning the stars or the wind; legal rulings have made it practically impossible to patent and own human genes. I mean inherently, constitutively, is there anything that cannot be owned because to do so would alter its identity; because owning it would change it into something else and negate the thing that it previously was? The chess piece called a rook cannot be moved diagonally and still be a rook. It then becomes another piece, a bishop, for example, or it reverts to simply being a piece of wood or plastic and is no longer a component of the world of the chess game at all.

Are human things like that? Does this shed light on why slavery is morally wrong? Ownership of a human being is incompatible with personhood; it redefines the slave into an object, not a subject, at least from the owner's point of view. An inappropriate move in chess simply changes the significance of an object that has only a contingent identity. A rook is a rook only within a game of chess; to set it aside as an object between games is not to do a moral wrong to it. But ownership of a slave erases the human subject, which is not a contingent identity but an inherent and necessary one. Setting a human being aside from the space of freedom and turning her thereby into an object only is a moral wrong to that human being. Are some nonhuman animals like that? Perhaps that is what wildness means.

Ownership as control and commodification as objectification for instrumental use are two sides of the same coin. To conceive of any good or service as a commodity involves three constitutive ideas: (1) commodities exist, and exist only, in the space of exchange relationships; (2) the value of commodities and exchange relationships is instrumental; and (3) commodities can only be privately or individually consumed.

A commodity is something that is traded or exchanged between distinct parties whose relationship is created by the exchange relationship. Commodities don't exist apart from distinct relationships based on exchange and, conversely, exchange relationships don't exist unless and until the objects or services involved in the transaction are understood as commodities. Commodities are not the same as gifts, and they are not the same as acts of love, fealty, honor, or service. I note these distinctions because these terms are appropriate in different kinds of contexts and relationships. Commodities are thus what may be called "institutional facts."[12] That is, they are potent social realities and not just illusions, but they only make sense within a particular social context.

The second key idea associated with the notion of a commodity is the idea that the good or service exchanged, and the exchange relationship itself, is of only extrinsic and instrumental value to the parties involved. The provider is only interested in the relationship with the consumer insofar as the consumer has something of value or worth (usually money) to the provider independent of the relationship. And the consumer is only interested in the relationship because it is only through that relationship that he can gain access to the commodity or service he wants or needs. The interests, wants, and needs of both parties exist prior to the relationship and the commodity exchange, they are instrumentally served (or not) by the relationship, and they will persist through time to create the need for future exchange relationships of the same type. In other words, the exchange relationship is something extrinsic to the self—one passes through it, uses it as a means to some other end, and then moves on.

Third, the notion of a commodity carries with it the notion of private or exclusive consumption. By that I mean that commodities are not simply used, they are used up. And when they are used and used up, this is understood to be an act of individuated appropriation. To the extent that you consume a commodity, someone else cannot. The concept of a commodity fits naturally in a situation of natural or artificial scarcity where resources are to be broken into discrete units and distributed across the members of a group or population for consumption. A commodity is something that ceases to exist as it is consumed.

We are now in a better position to return to Leopold's claim that private property, ownership, and commodification lead to abuse and a failure to care for natural systems.[13] There are some legal situations in which ownership of private property carries restrictions on its disposition and use, but for the most part to own something is to have the discretion to place it into a market of buying and selling and to use it as one sees fit, subject only to the general principle that it should not be used in a way that is harmful to others who do not consent to that harm. The concept of property in general describes a range of relationships between human persons and the material world. Private property describes a peculiar subset of such relationships in which humans have the right to commodify things and to use them up. Private property, framed by the social contract of consumption, is a bundle of individual interests and negative rights against interference by others in the way commodities are used to pursue subjectively defined desires. It is on this basis that we can understand what Leopold was getting at.

When we think we own something, it makes us careful about our own interests but shortsighted about the interests of others and completely blind to the notion that the object of property or the commodity itself might be said to have interests of its own. This often loosens the restraints we feel on how we use a commodity. As one farmer feuding with another might say, "I would shoot this dog before I would sell it to you." Moreover, it bases virtually all the restraints concerning property use on the self-interest or competitive advantage of the owner. If I own a commodity, I am at liberty to extract all the value I can from it as a material object or as an object of exchange. That is extractive power.

To what calculus is the exercise of extractive power subject? We have seen this before in the social compact of the ancient Greek Sophists. We abide by the common rules of the social contract of consumption, but our hearts are not in it, only our heads, only our rational self-interest. More often than not, those heads are afflicted with a kind of amnesia and dementia because this perception of self-interest is very short-term and confined to a limited set of metrics for their evaluation, usually defined in the artificial semiotic system that we call money. When Leopold says "we abuse the land," by "land" he means all of nature and living systems, and by "abuse" he means manipulation in a way that degrades biological and ecological function. He also means any use that is solely motivated by the individual desire and self-interest of the human being engaging in that use. For Leopold the land should not be exploited, farmed so as to extract the maximum yield of crops per acre, or manipulated to maximize financial profit at the expense of natural health and beauty over a time span longer than a single human life. Following this line of argument, let us say that the land should not be "used" but rather "worked." Recall Hannah Arendt's conception of "work" as the from-giving, meaning-making interaction between humans and their social and material worlds. In this sense, property ownership should not be practiced in the context of private consumption but rather in the context of public work. That is one of the key terms of the ecological social contract.

Viewing property rights and use as a kind of public work activity raises the question of what the notion of something being "public" or civic means, and it raises the issue of what

constitutes a commons or a common-pool resource in an ecological political economy.[14]

Public Work

Public work and commodity are different lenses for viewing
the economic activities of humans in nature. But then again, the
metaphor of a lens is too passive and static; it does not capture
the dynamic, constitutive quality of these conceptions. Better
to say that these conceptions are alternative interpretations of
practices and that the very act of reinterpretation by changing
the concepts we use has the power to transform these practices. To convince ourselves that the land is a commodity will
eventually turn it into a commodity because it will elicit the
cultural and social responses that commodities typically
elicit. This is not about naming, labeling, or describing; it is
about ways of living and ways of world-making.

The distinction I wish to draw can be introduced by an
example. Consider a simple example of something that typically would be understood as a commodity, and something that
most definitely would not. A bag of cookies that one must divide
among a group of eager children contains commodities—the
more one child gets, the less another will get. The cookies will
soon cease to exist as commodities because they will be consumed, and the relationship they have created among the
children (and the adult dispensing them) will cease to exist.

Now think about what is going on when one reads or tells
a story to children and then talks about it with them. Storytelling is a public work of understanding, not consumption. It
involves sharing meaning in common, not individualistically
appropriating it. Granted, it can be difficult to decide how to

draw the line between stories or meanings that should be considered proprietary and those that are part of the cultural or creative commons. But it is not difficult to see that there is such a line or to understand why. Meaning is inherently a common-pool resource because language (and other complex semiotic systems) is a commons. There is no such thing as private property in matters of language, as the saying goes. Aided no doubt by evolutionary neurological predispositions, a child learns language socially and environmentally by sensory exposure to, and the mind's interaction with, sounds, symbols, images, and meanings. Much like Rousseau's notion of inner legislation, grammatical form in language structures identity and relationality in place and time. Language is a commons like sunlight and air. Certain meanings can be quite intimate and personal, but meaning as such is inherently common and public.

Earlier I suggested what the constitutive features of a commodity are. Below are the parallel features, as I see them, of public work.

A public work does not create exchange relationships among the people taking part. It is a mutual sharing in and a joining with a larger dimension of being than any of the participants, acting alone and without the social practice, could attain by themselves.

Moreover, it is not an instrumental transaction or relationship; its purpose and value are not defined in terms of the satisfaction of preferences or quid pro quo.

Finally, while the supply of time, energy, or material goods involved in undertaking civic practices or public works might run low, nothing is being consumed in public work; it is being shared and dispensed. And it cannot run low because its plenitude or abundance is not diminished incrementally by those who draw upon it. Indeed, it grows the more it is partaken of, the more it is shared.

Public works establish relationships among individuals that are not transactional or consumptive but involve a co-operative and participatory effort to *produce* something of common value. This value is not appropriated exclusively by one of the parties to its creation, no one is simply a producer or simply a consumer, and the value is realized by communities as much as by individuals. In fact, a community is nothing more or less than a fabric of relationships formed by public work. That is why it is appropriate to call them *public* or civic works. Moreover, these cooperative practices are productive, not exchange, activities, and they require intentional, intelligent effort on the part of all those involved in order to produce something of public value or significance. Therefore it is appropriate to call them public *work*.[15]

What kind of communities will we have if we forget the ways in which mutual vulnerability and dependency can bring us together rather than drive us apart? Dependency and limitations can also be symbols of a common fate and a shared mortal humanity; rising to their moral challenge is public work of the most significant kind. A metaphor such as the notion of commodities obscures what is shared and accentuates what is individually and privately appropriated. I contend that this is always socially pernicious, but it is especially so today.

The Residual Public

We have inherited a group of concepts from classical Greek and Roman political thought that were developed in order to understand the human capacity for—and practical experience of—shared forms of life, purposes, and predicaments. Classical political theorists invented these concepts to capture their experience of the public that their own history, institutions, and environment made possible. These concepts (*polis, politeia, polites* in Greek; *civitas, res publica, cive* in Latin) are the roots of our own terms—"political," "public," "civic," "common," "community," "citizen," and "republic." We retain these terms as central to our political vocabulary, but their meaning can and has changed.[16] Similarly, we still have the cultural and institutional reality of the public within our own political experience; it remains a permanent possibility of the way in which human beings can organize themselves in large numbers, but it certainly is not a necessary feature of human society, and in the modern world its existence is fleeting and fugitive at best.

Shared purposes or problems are not the same as individual purposes or problems that happen to overlap for large numbers of people. Of course, they do affect persons as individuals and as members of smaller groups, but they also affect the constitution of a "people," a population of individuals as a structured social whole. An aggregation of individuals becomes a people, a public, a political community when it is capable of recognizing common purposes and problems in this way.

The social contract of consumption, with its liberal market political economy and its possessive individualistic or *homo economicus* conception of the self, has a truncated understanding of the public dimension of political, economic, and social life. It sees the public in individualistic and aggregative terms; it sees public things as "externalities"—things external to the market because they cannot not be effectively privatized and made subject to the calculus of individual self-interest; or it sees public things as goods or services that otherwise cannot practically or efficiently be produced and distributed in a for-profit mode of investment and pricing.

There are many available ways of defining and understanding the idea of the public. Within the consumptive political economy the main conception is an aggregative notion: public things are the collective sum of individual interests or states. For example, public health is thought of as population health or health status considered from a population rather than an individual perspective. The utilitarian thinking that is very influential in mainstream economics works with aggregative notions of gross domestic product and uses this comparatively to calculate economic growth. Also social justice and well-being are conceived in terms of aggregate net benefit across a population of persons. Indeed, this utilitarian approach has serious drawbacks in terms of distributive justice and equity, since an aggregate net benefit maximization is compatible with steep inequalities among the benefit shares that are received by certain individuals or groups. In fact, some recent uses of the social contract thought experiment in ethics and theories of justice, like that of John Rawls, have been explicitly designed to avoid this aggregative approach in favor of one based on intrinsic values such as fairness and individual rights. Whether the aggregative individualism of a welfarist theory of justice such as utilitarianism or the pluralistic individualism of a rights- or duty-based theory can leave us with an adequate concept of the public realm in the classical sense, or what a commons in nature and society is, remains to be seen.

Another concept of the public frequently used is the idea of the public as a space of external costs, or externalities. This space is a free dumping ground or free resource to be used, but, since it is not owned by any market competitor, it cannot be protected or managed except by political rules and authorities. The classic case is air or water pollution. The behavior of the polluter is not restrained by the market and profit motive because the costs of the pollution are transferred to someone or something else. They cannot easily be brought inside the balance sheet of the self-interested rational polluter. In fact, this external status gives the rational polluter a positive reason to pollute. In this way, the invisible hand of the market, its automatic, amoral mechanism, turns into an invisible fist. The valuable resources contained within the social or natural world outside the market sphere are important, such as the ecosystemic carrying capacity of a river to cleanse itself of toxins disruptive to its community of life. These are wealth in the sense of plenitude, or commonwealth, but are at risk of depletion and degradation. Ever in jeopardy, ever an orphan, the conservation of commonwealth calls forth public work or common action, *faute de mieux*, in the wake of market failure born of self-interest and an inadequate moral compass and moral brakes.

Different from externalities, another conception of the public developed by economists, public choice theorists, and others working out of the liberal welfarist and utilitarian tradition, resides in the concept of "public goods." Public goods, like external goods, are things that it is not easy or efficient to get private parties to pay for on a commodity basis. They benefit everyone to one degree or another, but no one acting in a private capacity can be said to be responsible for their care and maintenance. So they are vulnerable to overuse and under-support, the so-called free rider problem. National defense, the maintenance of wilderness areas, universal education, the arts, charitable assistance, natural monopolies caused by technology, such as communication bandwidth and now the Internet and the like, are often categorized as public goods in this sense.

In public health, the phenomenon of "herd immunity," the collective resistance to certain infectious diseases, of a population is another example of a public good. It can be undermined if a society fails to maintain an adequate vaccination program for children, even though there is but a small risk that any particular child will contract the disease in question and even though there is always, with any vaccine, a small statistical risk of adverse side effects. In taking steps through education or mandatory rules to maintain high levels of herd immunity, the public health authorities sustain a public good that the private sector, acting on particularistic self-interest, on its own would not be able to protect. Beyond self-interest, an ethic of parental duty and responsibility, if interpreted individualistically and focused on private children rather than on the public condition of all children, may not sustain herd immunity either. Here the solution can come with coercion by the state or it can come through the inculcation of a public-oriented or civic ethic. But then the protection of all our children, empathy and solidarity with strangers, would take us beyond the realm of public goods, as the social contract of consumption tends to understand them.

In each of these cases, that which is public or common is not directly understood on its own terms, according to its own logic of value or meaning, conviction or contentment. Instead the public is considered residual. It is that which falls through cracks of the primary driving force of the society—commodity exchange and for-profit market actions and motivations. In the consumptive social contract, public systems and values are expedient due to various shortcomings or pathologies of nature in humans—violence, personal failings that lead to dependence or social marginality, or an unjust disposition to take a free ride by benefiting from the productive and law-abiding self-restraint of others without undergoing that discipline or making a contribution oneself. These ways of understanding what is public rest, at best, on morally weak considerations of prudence and contingency. They are residual and transactional notions, while the ecological social contract offers what I would call a transformative conception of the public.[17]

Governing the Commons

Now let's look more closely at the notion of a commons, using the thought experiment called the tragedy of the commons.[18]

Imagine a large tract of rich grassland owned by no one and surrounded by several dairy farms. From time immemorial the herdsmen have allowed their cattle to graze some of the time on the common field. Since the grass there is free, whereas they must pay for the food their cattle eat on their own property, the herdsmen have an incentive to graze their herd on the common as much as possible. On the other hand, overgrazing will eventually destroy the fertility of the soil, and the ecosystem of the common will collapse, leaving less and less grass for everyone until eventually the land will become barren. So each farmer has at least a long-term incentive not to overburden the carrying capacity of the common.

If the motivational and behavioral assumptions of the consumptive social contract and its political economy are correct, the herdsmen will inevitably be driven to overgraze the land by short-term self-interest and fear of losing competitive advantage. The logic of self-interest and insecurity caused by the lack of a common ruler will doom the common, and hence be an irrational outcome as well as an environmentally harmful one. Of course, it could be that the herdsmen will be able to exercise self-restraint and will spontaneously refrain from taking advantage of the restraint of the other two. This would be a solution of virtue, not so much civic virtue perhaps, but altruism. Otherwise, between sainthood and disaster, there are three possible solutions.

1. *Create a common power.* One solution to prevent the tragedy of the commons is to solve the more general problem of order by creating a sovereign authority. This protects an environmental commons or a social commons beneath the shield of a political commons or a public authority to enforce collective restraints. This can be done by conquest of the strongest or by collective agreement through a social contract. In the first case, the commons would now become the private property of the conquering sovereign and would be protected as such. In the second case, it would become the common-pool resource of the whole society and would be governed as such.

2. *Transform the common into private property.* A second solution would be to somehow internalize the cost to the herdsmen for the use of the common resource. For example, an agreement could be made that would make the herdsmen pay for the grass when their cattle graze there rather than allowing them to exploit this resource for free. Or the commons could be subdivided and turned into the private property of each of the herdsmen in equal shares. But this is a paradoxical solution because it would save the commons ecologically by destroying it as a commons socially. It is interesting to note how this solution turns Leopold's idea on its head. Rather than encouraging careless use and abuse of the land, private ownership is seen by those who promote privatization as providing the motivation for good stewardship and land-use practices. However, this is not a direct disagreement between libertarian advocates of privatization and Leopold, so much as a result of the incommensurable concepts of value and wealth they and he are using.

3. *Protect the common (resource) by turning it into a common good democratically governed.* A third logical solution—the Rousseauian one, and today the appropriate ecological one, it seems to me—is to begin group deliberations to come to some kind of civic consensus on regulations and rules of using the commons without using it up. This would make its preservation and maintenance a matter of accepted duty and shared membership. Here the commons would be recognized for the first time not merely as a free resource but as a truly common or mutual good in a shared way of life; something that will benefit all the parties best and will be sustainable over time, if and only if none of the parties exploit it unduly or unfairly. Ostrom favors this approach and has been impressed by its practicality in the world today:

> Some scholarly articles about the "tragedy of the commons" recommend that "the state" control most natural resources to prevent their destruction; others recommend that privatizing those resources will resolve the problem. What one can observe in the world, however, is that neither the state nor market is uniformly successful in enabling individuals to sustain long-term, productive use of natural resource systems. Further, communities of individuals have relied on institutions resembling neither the state nor the market to govern some resource systems with reasonable degrees of success over periods of time.[19]

Now each herdsman, for the first time, takes on a dual identity. One is his private identity as a competitive entrepreneur and follower of short-term interests.[20] The other is a civic identity as citizen in the face of this new public thing. As a citizen he has not just private interests, but also public duties; he has not just the virtues of a successful businessman, but also civic virtues that call for the care, stewardship, and trusteeship of the ecological health of the commons. The third solution clarifies the distinction made above between transactional and transformative notions of the public. When property is defined from a public perspective as a common-pool resource, something that was seen as a raw material with only instrumental and human-centered use value comes to be seen as a natural gift and a communal trust. This is the perceptual shift in the gestalt of an ecological political economy. A private rabbit becomes a public duck.

How has the consumptive political economy in the liberal tradition during the fossil carbon era viewed these three solutions? It has rejected Hobbesian absolute sovereignty (at least within the developed world), but it is comfortable with a more limited version of collective authority, restrained by electoral competition among elites and the rule of law. The second solution, privatization of the commons, is the preferred solution of free-market liberalism even today. However, the third solution, which amounts to a more participatory and discursive democratic approach to the governance of common-pool resources, is gaining attention and has demonstrated its feasibility in certain areas.[21]

9

Freedom

Relational Interdependence

As noted earlier, our accelerating, global extractive assault on planetary resources and ecosystems, as well as the unprecedented extensions of our technologic reach, do not truly represent progress and the triumph of human freedom or the human destiny. Why not? For one thing, they are not sustainable or viable as a road to the future. No less important is the fact that technological advance and extractive assault contain an inner contradiction. While seeming to extend human freedom, they are laying the groundwork for its repression. Being at liberty to behave in ways that are ecologically irresponsible and destructive is not to be liberated; it is to be dominated by technology and desire.[1] Genetic engineering and biotechnology can be enthralling when the creative impulse to redesign overwhelms the creaturely instinct to accommodate, whether it be applied to agriculture, attempts at human genomic enhancement, or geo-engineering to mitigate climate change. So too it is with economic life. While seemingly representing the advanced expression of human capability, technological advance and extractive assault are actually undermining what is most precious in humanness.

I believe that these questions are bound up inextricably with the future of freedom. The rediscovery of the truly creative and the inescapably creaturely dimensions of our human condition requires the transformation of the current neoliberal world of extractive liberty and possessive individualism into a world of relational freedom exercised through the practices of justice or parity of social membership and solidarity.[2] We don't quite know yet how to foster the psychosocial development of such ecological selves or write their collective biography on a large scale. But we desperately need to learn.

Thus the concept of liberty or freedom (I use the terms interchangeably) has a central place in the formation of a new ecological social contract and its political economy.[3] I seek to discover which particular conception of freedom is most consonant with the circumstances of an ecological political economy. My critical goal is to scrutinize received conceptions of freedom that are so individualistic or atomistic that they cannot realistically and reasonably guide human norms and self-understandings in an ecological era. My constructive goal is to formulate an understanding of freedom that is consonant with the demands of such an era. What is needed, I believe, is a theory and practice of "relational freedom."[4]

Consumptive Freedom

A fateful hallmark of the modern era is that we have based our political economy, our law and governance, and much of our moral philosophy upon a distorted and self-defeating understanding of freedom. This notion of freedom as the individualistic pursuit of limitless need and desire, which is often called "negative liberty," is fully consonant with our blindness concerning our true ontological place in nature.

Liberty is often used to designate a condition (or a potentiality) of mind and agency that inheres in individuals as a matter of right. Specifically, freedom understood as negative liberty is *self-directed individual agency free from interference by others in the pursuit of the subjective satisfaction of needs and desires that are ontologically and psychologically limitless and insatiable.*[5] The right to freedom so understood is a moral claim that can be made by a person against others who are in a position to impede, impel, or coerce the person's behavior in ways that conflict with the person's own purposes or interests. Private individuals may fall into this category, but the right to be free quintessentially applies against those who wield corporate institutional power or the legal police power of the state. The most potent source of the individual's security and protection can also be the most dangerous threat to his or her freedom. This is a paradox that liberalism has struggled to solve, yet it is also a paradox of liberalism's own creation.

The consumptive social contract and neoliberal political economy esteem individual freedom very highly, often above all else. It is ambivalent, however, about how far that right to liberty extends. One interpretation holds that the freedom of the individual creates only a negative obligation of forbearance (noninterference) on the part of others. A contrasting interpretation maintains that the individual right to liberty sets up a positive obligation of action by others to assist the individual in obtaining the resources and capacities that will make his negative freedom meaningful. This is the distinction, roughly, between libertarians and market liberals, on the one hand, and egalitarians and social liberals, on the other.

These nuances, while very important, are not at the root of the problem today. A more fundamental question for an ecological political economy concerns the moral justification for limiting or overriding the individual moral claim to noninterference or to "self-sovereignty." John Stuart Mill argued in *On Liberty* that such a justification can only be provided by the condition of involuntary harm to other human beings. In practice, that criterion gives individuals very wide latitude in their interactions with nature and is difficult to reconcile

with an ecological perspective on the social and cultural consequences of individual agency and institutional or technological
power. One way out of the dead end of libertarian conceptions
of freedom is to assign other principles or values a higher moral
priority than freedom, and subordinate or override freedom
when it comes into conflict with those values. Another approach,
which I myself favor and shall explore here, is to seek within
the logic and meaning of the concept of liberty itself internal
self-limiting conditions. Does the concept of individual freedom, in anything like the form that we have inherited it from
the liberal political tradition, contain within itself the basis
for its own moral limitation?

For economic liberalism, freedom rests upon subjectively
defined value. This is because each person is held to be the
most reasonable custodian and definer of his or her own interests and objectives. If the power to determine those interests
is exercised by others, especially by officials of the state, a
person is deprived of freedom and is hampered in the development of independent-mindedness, skill, and self-reliance.
For his part, John Stuart Mill considered these capacities to
be some of the hallmarks of human flourishing, and later
economic liberals have viewed them as essential for economic
efficiency and growth under capitalism.[6]

Moreover, for economic liberalism there is no independent
standard of reason to determine if one person's use of freedom
is inherently superior to another's. I noted this earlier in the
discussion of determining a "just price." There is no intrinsic
right or intrinsic wrong: the free person is not subject to such
elitist and arbitrary value judgments imposed from above or
outside. In the modern ideology of neoliberal capitalist political economics, individual need or desire to acquire and consume is taken to be psychologically and ethically unlimited.[7]
Absent tangible and serious harm to others resulting from an

action, individuals *should* be allowed to make their own choices and life plans. If individuals are permitted by social and political arrangements to have this freedom, mainstream liberalism has argued, the society as a whole will prosper, and the result will be a well-ordered, but not stiflingly conventional, life of freedom.

On the contrary, I maintain that when freedom is based on self-defining or market-defined utility alone, the distinction between liberty and license withers away, and freedom ceases to be answerable to anything higher than itself.[8] Preventing harm, at least in a fairly narrow sense in which liberalism has defined it, does not constitute the entire range of ethics in individual life or in public policy and governance. The absence of harm does not entail the absence of wrong or injustice. Protection from harm may preserve and promote life, but it is not sufficient for achieving a good life. No, we should morally charge neoliberal political economy on several counts besides harm to other people or even harm to the planet. More positively, we should also stress the promise of another freedom, a freedom over and above the negative right to be left alone, a *relational* freedom inherent within an ecological way of structuring human productive economic agency and governing it.

Relational Freedom

Negative liberty (freedom from) and positive liberty (freedom to) are two main senses of freedom. Relational liberty is freedom in a third and distinct sense; relational liberty is freedom in and through relationships of interdependency.[9] More specifically, my notion of relational freedom can be defined as *freedom through transactions and relationships of interdependency with others that exemplify justice (parity of social membership, voice, and participation) and solidarity (mutual-ity of civic concern and respect).* The essence of the philosophi-cal strategy I propose is to internalize the common good into the individual good, to read the "We" into each "I."[10] And my approach is to internalize the freedom and well-being of all (both human and nonhuman creatures and systems of life) into the freedom and well-being of each.

Only in this way, I believe, can we break out of the cultural trap and impasse of pitting individual freedom and collective restraint and limitation in opposition to one another. Only in this way can we counter the moral license that liberalism has given to self-centered liberty and insatiable desire—market-oriented economic progress and growth—in its conception of the human good. I believe this will provide an essential lynch-pin for an ecological social contract.

Relational freedom resists the nihilistic "creative destruc-tion" (economist Joseph Schumpeter's apt and candid phrase) that the neoliberal political economy has institutionalized and that is so socially disruptive and environmentally rapa-cious.[11] Relational liberty, freedom through interdependency, is a warrant to live one's own life in one's own way that results from embedding that way of life in a tradition, a civic life of shared purpose, and rooting that life in a sense of ecological place and in a sensibility of care for place and care for earth's life-support systems. Just as there are certain kinds of practice or activity that by their very nature cannot be done alone, so there is a kind of freedom that subsists not in separation from others but through connection with them: not in pro-tections but in pacts of association; not in closed futures but in open ones; not in fences but in circles; not in extraction but in conserving; not in artificing but in accommodating.

Relational freedom rejects two constitutive features that have characterized theories of freedom in the liberal tradition.

One is the privileging of individualistic values over communal ones—individual freedom trumps community solidarity. The other is setting up a conflict or antithesis between the individual and the community in the first place.

These two features make liberal theories of freedom remarkably devoid of the web of interdependencies—that is, culturally meaningful roles, styles, and self-identities; shared values, rituals, and practices. These theories, particularly when applied to the realm of economic agency, tend to portray an idiotic world of atomistic individuals, each with their own self-regarding interests and life plans. This requires, at most, a social existence of peaceful and predictable transactions for mutual advantage. A thicker sort of mutuality based on care, empathy, friendship, or solidarity is rarely necessary, and from a liberal perspective always epiphenomenal. Furthermore, it is this feature of atomistic abstraction that makes such philosophical accounts ideologically functional for a political economy founded on the logic of extractive power. Interestingly, and paradoxically, atomistic individualism both conceals and normatively justifies extractive power at the same time.

Again, on a relational conception of liberty, freedom is constituted, not in spite of connections and convictions—membership and mutuality—linking self to others, but in and through these connections and convictions. Enacting relational freedom in one's life develops a self-identity built out of ongoing practices that exemplify the creative and aesthetic dimensions of a humanity naturally flourishing. Such relationships involve not only social interactions with other human beings, but also—and crucially—meaningful connections between the person and his or her own activity on and in the world of fellow creatures, living systems, and material objects. A relational conception of freedom and of the person contains a counter-vision to notions of alienation, commodification, and the objectification of the human or the natural Other. Such a vision leads away from the control of natural material as a source of "wealth" defined as material accumulation, relative social status, and utility maximization. It leads toward a notion of artistry, craftsmanship, and the accommodation of the inherent properties of natural form.[12]

Planetary thermodynamic processes may be systems that operate impersonally without a locus of intention, control, or responsibility, but economies, societies, and political communities are not systems in that sense but rather are structures of purposive human agency. However, adopting an agent-centered rather than a systems theoretic approach, as I have throughout this book, does not entail an ontological or atomistic individualism. Human acts are intentional, purposive, and meaningful both to the actors and to others who share in the rule-governed forms of life and communication within a society and culture. The ethical norms that fit into human agency therefore are not limited to self-referential states of interest or desire. In order to understand ethical conduct—or in order to engage in ethical discourses of justification and other forms of argument—one must have recourse to concepts and categories that reflect the relational nature of the human self and the contextual, socially, and symbolically mediated nature of the self's interactions with others.[13]

The implication of this is quite important.[14] An ecological political economy will come about only through change at both the level of individual behavior and of social norms and institutions. In practice this means that we learn to articulate the values and ideals that the members of these societies would express if they thought and acted like interdependent and relational selves—ecological selves, or ecological citizens and trustees. Part of the task of an ecological social contract is to shape this self-identity and foster a moral imagination that can see the good and freedom in relational terms. Mainstream economic activity over the years has helped to build a population of possessive individualists through its doctrines and through the institutions it has legitimated. Today, we must undertake the task of educating a new generation of social persons. The time has come for economic knowledge and discourse to show all of us, specialists and ordinary citizens alike, that our personal flourishing is inextricably linked to the flourishing of others and to the flourishing of the natural world. An ecological political economy does not abandon the concept of economic self-interest, it supplements it, and, even more important, it transforms it.

For this task, the ecological social contract needs the vocabulary of solidarity, mutuality, reciprocity, community, and the common good. The pioneering thinkers in the fields of ecological economics, steady state, and degrowth have already begun to develop this vocabulary.[15] Beyond the notion of moral obligations that are correlative to the individual rights and interests of others and the obligation to do no harm, ecological political economy also needs to appeal to a motivational structure that is informed by ecological citizenship and trusteeship. As I shall argue in Chapter 10, ecological citizenship is not so much a role, or a legal status, as it is an orientation and a disposition of living that is grounded in a sense of responsibility for promoting and sustaining the common good of the community as a whole and its natural context.[16] People who have this sense are plain members and citizens of the biotic community. If the people around the world, and especially in those nations whose economic behavior has to change most drastically, have lost the capacity to comprehend these ideas, then it will not be possible to either coerce or empower them to undertake the kind of collective institutional and behavioral change that creating an ecological political economy will require.[17]

The Moral Content of Right Relationship and Relational Freedom

Not just any form of human interaction or transaction should count as a relationship through which the value of human freedom is constituted. Interactions of domination, exploitation, coercion, violence, seduction, or duplicity, each of which effectively reduces human beings from the conditions of subjects to the conditions of objects, do not count as "relationships" that constitute freedom. A rapist is not more free than a loving and law-abiding person; quite the contrary. When it comes to relations with nonhuman beings, such interactions may be ruled out ethically for the same reasons insofar as the nonhuman organisms possess agency—as clearly many animal species do—and thus are subjects to be respected and not simply objects to be used.[18] Nonhuman beings that cannot be understood as purposive agents are still subject to norms and restraints on human action that follow from intelligent, experience-based and evidence-based recognition of ontological interdependence within the web of life and existence.

In short, neither in relationship with other subjects, nor when using natural objects, are human beings completely at liberty to do what they will. The principles of right relationship and right recognition denote the dimension of freedom that promotes a kind of moral (and political or civic) community that is open rather than closed—a hospitable community of participation, voice, mutual care, and solidarity. Hence, at the heart of the concept of liberty is the notion that to be free is to be in certain kinds of relationships with others. Human freedom in this relational sense cannot be understood in isolation from either the social *or* the ecological biotic webs within which the individual resides and are the preconditions of human survival, development, and flourishing. Relational freedom cannot exist within the context of unjust structures of power, wealth, social opportunity, health, and psychological integrity. Neither can effective human economic activity exist amid the degradation and breakdown of biophysical systems. This provides a criterion for evaluating which types of relationships (transactions/interactions) are to be nurtured, facilitated, and promoted by common rules and public policy, and which are to be discouraged or prohibited.

Today the chasm is growing between the vision of relational freedom in an ecological society and political economy and the worldview of both laborers and consumers in the contemporary West—and indeed throughout much of the developing world as well. The everyday lives of individuals are permeated by mass consumer culture. The vocabularies that people use to define a self-identity and to comprehend their situation are growing increasingly thin and impoverished from both an ecological and an ethical point of view. People with a consumer's sense of relationship and a tourist's sense of landscape, ecosystem, or natural place cannot grasp that our well-being depends on healthy natural and social systems; that we have responsibility for preserving and restoring them; or that our freedom is threatened by those very same institutions and practices that undermine the natural world.

The experience of relationship and place must be altered and enriched—from consumers to citizen trustees; from tourists to inhabitants and stewards. If it is so altered, ecological political economy will flow out of that altered lifeworld and experience. But isn't the alteration of that ordinary experience—a transformation of the sense of self and the motivational bases of moral imagination—itself something that can only come about in the context of an already established ecological orientation in economics and politics? Certainly the conceptual resources for an ecological social contract are available for us to embrace and develop. But, even if we can muster the collective will and courage to do that, can we do it in time?

10

Citizenship

From Electoral Consumer to Ecological Trustee

If new forms of renewable energy can be made relatively inexpensive, then the new social contract of the future may continue to be something like the social contract of consumption. That could spare us some of the difficult social transformation I envision, but such a technological fix could also spell disaster for biodiversity and ecological resilience, quite apart from the problem of climate change. However, more likely in fact is the scenario of much more expensive energy, a lower material standard of living, and a curtailed consumptive pattern of behavior for that portion of the world's population who are now the most affluent.

If the fossil fuel era is coming to an end for natural reasons beyond our control, then the social contract of consumption will have to give way to an ecological social contract to provide a new constellation of legitimacy, a new foundation for both contentment and conviction, a new basis for political and social stability. On the basis of the ecological contract we can build a consensus and political will to move from the affluence of consumption to the plenitude of conservation, from using up to sustainably using.

It remains to be seen whether governance under an ecological social contract and in an ecological political economy will take on an authoritarian or a democratic form. But either way, that governance will be substantially different from the present governance system of liberal representative democracy and growth-centered political economy.[1]

Can we devise a way to make the governance that follows from the new ecological contract democratic in a thick and robust sense rather than authoritarian? Can our sustainable political future go with Rousseau and be governed by ecological democratic citizens, or must we go with Hobbes and be governed by a unitary sovereign power: an ecologist king in an ecological Leviathan state?

The authoritarian version of the ecological contract is a relatively autocratic or oligarchic form in which the citizenry gives authority and obedience to a ruling elite in return for ecological security; protection from the horrors of catastrophe and collapse. The democratic version is a more discursive and directly participatory form of a constitutional democratic republic in which a consensus-building process from the bottom up sets general principles and goals, while professionals and other expert elites are given the technical function of determining laws and regulations (and the police power to enforce them) to secure ecological integrity, sustainability, and security. In both versions a new basis of legitimacy and of contentment and conviction come, not from material hedonism, but from a communally and relationally rich quality of life. As Hobbes and Rousseau well understood, the authoritarian covenant of ecological trusteeship can be sustained by fear and self-interest alone, but for how long? The democratic form of this covenant requires something much more difficult but perhaps more lasting—it requires a transvaluation of values: an expansion of the moral imagination that is ecological citizenship and civic virtue, and perhaps an ecological civic faith or fidelity as well.

Either by fear and the desire for security or by principled normative conviction and an internalized sense of what is

good and fulfilling in life—one way or the other, this new ecological contract will provide legitimacy and stability even in
the face of much less material affluence than the old social
contract of consumption has provided in the fading halcyon
days of cheap carbon energy.

Liberal representative democracy or interest group democracy is a form of democratic governance prone to incrementalism and preservation of the status quo—or at least great
continuity of expectations and practices. Does the time scale
and the response horizon of interest group democracy fit with
the time dimension of the current ecological crisis? Can it produce policies and social changes that will reduce greenhouse
gas (GHG) levels below 350 parts per million by 2030? If not,
what form of governance will produce the necessary public
will and compliance? What type of new institutional structure
for governance (not simply government itself, but also the pen-
umbral process of the political culture, the news media, and
the civil society) will be able to act in the thoroughgoing and
relatively rapid ways necessary?[2] What kind of new ecological
citizenship will this ultimately require? I shall discuss in more
detail some of the options of ecological governance in Chapter 12. Here I focus on ecological citizenship.

Ecological Citizenship

What is ecological citizenship? The distinction made by the
philosopher Charles Taylor between an "opportunity concept"
and an "enactment concept" is helpful here.[3] Citizenship is *not*
best understood as a bundle of rights that may or may not be
exercised at the personal discretion of the individual (such as
the right to vote). This is to see it in terms of opportunity.
Instead, citizenship is best understood as an active freedom

that involves a particular set of practices, forums, skills, and opportunities. This is to see it in terms of enactment. Citizenship is not like a commodity or possession one owns and uses (or not); citizenship is a form of life, a kind of social being that one can cultivate and pursue. In an authoritarian society the opportunities to cultivate that form of life may be severely truncated or even nonexistent. In a democracy these opportunities must be open as a matter of right to virtually all adults.

For the purposes of ecological citizenship, it may well be that familiar approaches in environmental education, such as personal instruction and information sharing, will not suffice. And the movement in the European Union to enhance democratic environmental governance by creating an enforceable right of access to information about environmental effects of governmental and corporate activities may only be a first step.[4] Ecological citizenship may require the creation of active "publics" that seek out knowledge in the process of engaging in civic action as well as the provision of information to individuals.[5] Ecological citizenship may require community organizing and the deliberate creation of enhanced "social capital" or civic renewal no less than it requires the services of trained professionals to provide technical knowledge, counseling, and guidance.

Let's consider the concept of citizenship in more detail. I think it is important to distinguish between what I will call *the citizen as rational-choice consumer* and *the citizen as deliberative trustee*. The basic distinction here is between: (1) a form of political and social behavior that involves the calculation of individual self-interest or group interest and the creation of a strategy to devise the most rational means to protect and fulfill those interests; and (2) a form of political and civic behavior that involves deliberation, either in a group setting or as a solitary individual, to orient oneself to the common good.

One basic tenet of the social contract framework is that the moral authority of government rests on the active, informed consent of the governed. What all varieties of social contract have in common is the value of respect for persons as free and equal agents capable of making promises, with all that capability entails, and the value of being a member of a cooperative of mutual care and respect. In the democratic version of a social contract, the legitimate exercise of power must rest on the consent of the governed because ultimately no one knows better than the governed what is in their own best interest and in the common good. All voices and all value orientations should be heard. No competent adult should be excluded from the practice of consent if he or she is willing to assume the responsibilities of membership or citizenship; the assumptions of natural superiority and hierarchy that accompany nondemocratic ideologies are absent in democracy. A basic faith in the intelligence and perceptiveness of the common person pervades democratic thinking. I use the term "faith" advisedly here. Discursive and participatory democracy requires both empirical reason and transcendental imagination, both transactional and transformative relationships, both enlightenment and enchantment. In his intellectual biography of John Dewey, Robert Westbrook writes:

> Dewey was the most important advocate of participatory democracy, that is, of the belief that democracy as an ethical ideal calls upon men and women to build communities in which the necessary opportunities and resources are available for every individual to realize fully his or her particular capacities and powers through participation in political, social, and cultural life. This ideal rested on a "faith in the capacity of human beings for intelligent judgment and action if

proper conditions are furnished," . . . It was this ideal and this faith that set Dewey apart from the mainstream of American liberalism. In terms of the criteria of Dewey's testament, much of the history of modern American liberal-democratic theory is a history of treachery, for a rejection of Dewey's democratic faith has become a standard feature of the dominant strain of liberal democratic ideology. Since early in the century most liberal social theorists in this country have regarded this participatory ideal as hopelessly utopian and potentially threatening to social stability. Unwilling to abandon democracy altogether (although there are some notable exceptions), these liberals have argued, in the name of realism, for revised and more limited conceptions of its ideals. Politically, many have come to favor Joseph Schumpeter's famous definition of democracy as "that institutional arrangement for arriving at political decisions in which individuals acquire the power to decide by means of a competitive struggle for the people's vote," a definition that narrows democracy to little more than an ex post facto check on the power of elites, an act of occasional political consumption affording a choice among a limited range of well-packaged aspirants to office.[6]

Democratic notions of right relationship and right recognition can also have a positive effect on commitment to fundamental obligations of humans in nature and toward nonhuman species and living systems.[7] From a democratic perspective, laws and public policies are ethically justified and legitimate to the extent that they emerge from the reasonable deliberation of free and equal citizens who will be significantly affected by them.

It is important to note that, while it has been consonant with liberal individualism, the consumptive social contract has not really been democratic. It has the form of representative democracy but rarely much discursive and participatory substance. It trades public responsibility off for private material self-interest and elite protection. Thus the understanding of citizenship has been privatized during the era of the social contract of consumption, and the deliberative element has been narrowed down to a representative (in theory) elite. However, as it is practiced by legislative and administrative elites today, the very notion of *deliberation* has been reconceptualized and transformed into a discursive practice that is quite different—namely, *bargaining*.

Liberal representative democracy thus pitches citizenship at two levels: first, for large numbers of people, citizenship is the right or opportunity to select representatives. Second, for a much smaller number of activists and organized interest groups, citizenship is active engagement in the political bargaining process—that is, a system of interest-group negotiations to determine the distribution of various kinds of resources, tangible and symbolic. The governance challenge ahead resides in the fact that this type of democracy is no longer sufficient. An adequate ecological governance, if it is to remain democratic in any meaningful sense, will require reinvigorating the practice of deliberative and discursive democratic citizenship among both ordinary citizens and among activists and elites. Only this will provide an alternative to rational-choice consumerism (of candidates or issues) and to the politics of bargaining among representative elites and interest groups.

Rational-choice consumerism manifests itself most obviously and directly in our private lives as consumers of both economic and political commodities that we perceive meet our private interests—whether it is the purchase of a new car or the passage of a policy of governmental deregulation and tax cuts for the top quintile of the population. As the latter examples indicate, consumerism cuts across the economic/governmental or private/public boundary line. Rational-choice consumerism affects how we interact with each other in political communities and how we attempt to influence government on matters of policy and law. Indeed it is the conventional wisdom in political science and in practical politics that this self-interested "What's in it for me?" approach is the necessary and proper orientation to take.[8] The rational-choice consumer and the "private citizen" become one and the same. All politics becomes first-person singular. In this, the privatization predominates; the civic dimension of citizenship gets erased.

Ecological citizenship depends on the capacity to see and to make connections. Civic deliberation feeds on the imaginative capacity to see beyond the limits of one's own situation and experience. Publics or communities are formed when a significant number of people develop that capacity and orient it in the same direction. To form a public is thus quite different from creating an interest group. A public is constituted by a perception of a shared or common good, not by a strategic alliance based on overlapping private interests.

The medium through which this perception of the common good of both human and biotic communities arises may take several forms. It is founded on shared or widespread experiences of a certain kind, such as the experience of struggling to gain recognition and respect for nonhuman nature in a stressed and overextended local economic system. Such experiences are then filtered through existing forms and patterns of cultural meaning and collective understanding. This interpretive activity takes place at all levels and fills the interstices of a neighborhood's or an ethnic community's life. It is at work in conversations among women shopping at the market, and men on lunch breaks or in social gatherings. It is at work in houses of worship and service clubs. It is at work in political meetings or other kinds of civic assembly.

Finally, these shared experiences form the basis for what might be called public judgments by being discussed and shared with other members of the community through a participatory process of deliberation.[9] In deliberation, the ordinary discourse of storytelling—more precisely, the attempt to make sense of what is happening by assimilating it to familiar cultural paradigms—is focused by the exchange of reasons and justifications for one's position. It also involves a concerted attempt to assess the significance of what is going on and, if deemed appropriate, to take some kind of collective action in response to the problem. Judgment and deliberation are activities of democratic citizenship par excellence. They build and exercise the moral and the civic imagination.

Rethinking Self-Interest

The two orientations of discursive citizen and rational-choice consumer can also be contrasted in terms of the basic social and psychological orientation that informs them. Discursive citizenship is essentially a dialogic, collective activity, while the consumer orientation is essentially monologic and solitary. When one deliberates, one engages in a dialogue of arguments and counterarguments, reasons and counter-reasons, with others. This dialogic character of deliberation is obvious when it is done in a town meeting or at some other community gathering; it may not be quite so obvious, but still remains the case even when an individual is alone thinking through a problem. An interior dialogue takes place in the person's mind as he or she imaginatively reconstructs the give-and-take with fellow citizens in a group setting. A decision reached through the dialogue of deliberation is not a personal or individual decision, strictly speaking, although each individual may share in the decision. It is a collective or common decision in the sense that it grows out of a process that has revealed a common good and a common resolve.

With decisions to consume, by contrast, the individual consults his or her own interior preferences, desires, goals, and personal values and makes a decision based on the principle of realizing one's own self-interest. The give-and-take of reasons does not assume the form of a dialogue in this case because even if others have given you their opinion or their advice about what to decide (buy), those views are treated as external information (advisory opinions) only. The ultimate decision will be an individual decision taken by the individual alone as the final best judge of his or her own values and preferences.

This is true no matter what the object being "consumed." It could be a product or commodity. Or it could be a candidate's platform, character, or views that are made the object of choice and consumption by casting a vote for the candidate at election time, or deciding to donate money to a campaign, or even a decision to actively make phone calls or hand out leaflets. All of these are acts of citizenship as a form of consumption and consumerism, rather than citizenship as a practice of deliberation.

The point is not to eliminate private or consumptive citizenship. Nor do I mean to denigrate it. The point is to supplement and leaven it democratically in Dewey's sense. Moreover, it is not my argument that self-interested orientations should be entirely purged in human social life. I doubt that it would be possible and, if some past attempts at "thought reform," and "reeducation" are any guide, it would not be desirable. Individual autonomy and the freedom to support policies that benefit your interests are long-standing values not only of the liberal representative democratic tradition, but also of the participatory democratic tradition. Social revolutions that have made a concerted attempt to eliminate consumerism and self-interest from political (and even private) life have ended by betraying democracy and imposing frightful forms of dictatorial and totalitarian rule. The problem is not the presence of self-interest in politics; the problem arises when *only* consumerism and self-interest are present.

Can the notion of individual human interests be salvaged in a post–fossil carbon age by redefining it? Can it come to be identified with sustainable modes of living? Or will the post–fossil carbon age also be a post-liberal and post-individualistic age in which the concept of interests is not merely transformed and redefined but actually overridden and subordinated to a marginal place in our moral and political lives? It is overweening at present, to be sure, but for how much longer?

At least we can say this much: It is unrealistic to expect that the virtues of deliberation and an orientation toward the common good will be the natural starting point for most of the people who come to democratic community meetings and who keep coming and stay involved. By and large, the consumerist orientation is going to be very strong—if not dominant—at the grassroots level, at least at the outset. People will invest their time in such a process only if they feel that they will benefit from it and that it will serve their interest. This is particularly true of a minority community that may feel especially disenfranchised, marginalized, and alienated from the mainstream political system and civil society.

If the main challenge of authoritarian rule is exercising power without unsustainable coercion, the challenge facing the deliberative side of citizenship is how to create its spark—its dialogic civic virtue—in the first place, and how to develop and reinforce it over time. What are the kinds of institutional settings and structures that will lead a group of people naturally and normally out of the consumerist stance and into a mode of deliberation? Out of monologue and into dialogue? It is hard to get anyone to participate in much of anything these days; significant barriers of time, mistrust, and hopelessness must be overcome. But it is probably easier to motivate people when you are able reasonably to appeal to their interests than it is to promise them the very hard work of coming to think, see, and imagine in new ways.

Yet, if I am correct in thinking that the future governance of an ecological political economy will need a sense of the common good and dialogic interaction, this is precisely what ecological citizenship asks of us. Only thus can we manage the transition from the social contract of consumption to the ecological social contract successfully.

Part IV

The Political Economy of Climate Change—Democracy, If We Can Keep It

This new society . . . has only just begun to come into being. Time has not yet shaped its definite form. The great revolution which brought it about is still continuing, and of all that is taking place in our day, it is almost impossible to judge what will vanish with the revolution itself and what will survive thereafter. The world which is arising is still half buried in the ruins of the world falling into decay and in the vast confusion of all human affairs at present, no one can know which of the old institutions and former mores will continue to hold up their heads and which will in the end go under. . . . Working back through the centuries to the remotest antiquity, I see nothing at all similar to what is taking place before our eyes. The past throws no light on the future, and the spirit of man walks through the night.

—Alexis de Tocqueville[1]

Ecological culture cannot be reduced to a series of urgent and partial responses to the immediate problems of pollution, environment decay and the depletion of natural resources. There needs to be a distinctive way of looking at things, a way of thinking, policies, an educational programme, a lifestyle and a spirituality which together generate resistance to the assault of the technocratic paradigm. Otherwise, even the best ecological initiatives can find themselves caught up in the same globalized logic. To seek only a technical remedy to each environmental problem which comes up is to separate what is in reality interconnected and to mask the true and deepest problems of the global system.

—Pope Francis[2]

11

The Ecological Contract and Climate Change

The news from Denmark at the conclusion of the 2009 Copenhagen Climate Change Conference was not encouraging. *Hamlet* was performed on the world's stage, and several years later the tragedy continues. The king's claim to the throne rests on dubious moral foundations. The prince vacillates, desperately trying to ferret out the truth before he acts, seeking political resolve in epistemological certainty. Courtiers jockey for position and advantage. Meanwhile the people and the land groan under the weight of a disordered state and injustice. Catastrophe and barbarism are massing forces near the border.

Shakespeare's timeless story may turn out to be, one hopes, too dark a lens through which to view the world's response to climate change, but the tragic vision is tempting, nonetheless, and hits close to home on all too many points. As we witness the current struggles by global princes to respond to the pronouncements of scientists ("Are they honest ghosts?") and to set meaningful limits to unsustainable economic forces and interests, indecision so deep-seated that it amounts to a paralysis of political and moral will is a gathering darkness. Hamlets are at the helms throughout the world. The timetable of the challenges facing us and the timetable of our collective capacity to respond are tragically out of joint.

Indeed, in the years since the debacle of the Copenhagen meeting the international process to bring greenhouse gas

(GHG) emissions under control and avert massively disruptive global climate change has struggled to regroup and gain practical momentum. In the meantime, scientific evidence has accumulated that makes the rapid pace and widening extent of atmospheric warming and changes in the world's oceans ever more alarming. Indeed, serious effects of climate change had already begun to be felt in some parts of the world well before 2009, but then one still might talk about averting worse effects yet to come. Today the discussion has shifted from averting to adapting and making ourselves resilient in the face of the assaults of an impending new planetary climate.

In view of this, surely it is reasonable and prudent to begin with the premise that eventually—soon, within two generations at most—natural limits will require radical transformations in human institutional structures, cultural traditions, and individual behavior. These transformations will affect the freedom and material standard of consumption of billions of people in the most powerful countries (economically and militarily) in the world today. Some of the states involved have authoritarian and elite-controlled governance structures; others are more liberal and democratic. But no government in the highly carbon-consumptive parts of the world can be indifferent to the conditions of its own normative legitimacy among its own people; no government can simply ignore the direct interests and needs of at least its middle and working classes. No matter how steep its sides or how narrow its pinnacle, each pyramid of power and wealth rests upon its base.

One wonders if the most carbon-intensive nations have the resolve and the governance capability even to adapt effectively. Much critical political and moral argument will be needed to spur more resolute and successful global action. Each nation and political culture in the world needs to take a hard look at its own internal dynamics and its fundamental convictions regarding ecological governance. The forces blocking an action consensus among nations lie within nations, at least to a significant extent.

In the immediate aftermath of the Copenhagen meeting, an analysis by *New York Times* reporter John M. Broder provided insight for that task within the United States. For good or ill, the American role in determining how successfully the world can avert climate change catastrophe will be huge. But the political commitment to exercise this leadership and to pave the way for other countries to act does not exist in Washington thus far. The worst outcome of Copenhagen is arguably inaction, but close behind would have been agreement on a new international treaty with teeth. Why? Because the U.S. Senate would not ratify it, just as it would not ratify Kyoto in 1997. In 2009 Broder wrote: "The Senate is split on global warming policy into numerous factions divided by ideology, geography and economic interest. And that's just the Democratic caucus. Republicans are nearly united in opposition to the kind of legislation that would be needed to match Mr. Obama's ambitions [reducing emissions 17 percent below 2005 levels by 2020]. Without Senate action . . . Mr. Obama's promises are merely that, almost certainly not enough to persuade other nations to commit to greenhouse gas reductions."[1]

One important rationale for unilateral executive action was provided at the time by the finding of the Environmental Protection Agency (EPA) that greenhouse gases pose a threat to human health and welfare. Again Broder comments: "E.P.A. regulation is the trump card that the administration is holding if Congress continues to dither. But Mr. Obama has repeatedly said that he much prefers a messy Congressional compromise. Trying to remake much of the economy by regulatory fiat is certain to become entangled in years of litigation."

In 2008, candidate Obama said of climate change: "Delay is no longer an option. Denial is no longer an acceptable response." Broder concludes by noting that Copenhagen would end with an interim political deal to keep talking about a binding treaty next year, and that is just what did happen. "Delay, it turns out," he remarks dryly, "was the only option."

Of course, delay is not really an option at all, but an abdication of ethically responsible governance. Although in the intervening years President Obama has taken some executive actions regarding climate change and greenhouse gas emissions, a Republican majority now controls the entire U.S. Congress, and the prospects for any kind of concerted and unified alternative energy policy or carbon tax scheme in American governance seem more dim than ever. How can the requisite conditions of political and moral will be mobilized so that a new sustainable global political economy will emerge in time? Can the kind of interest group liberalism that Broder describes respond to this challenge successfully? And, more sobering still, can any type of democratic governance do so? Can the progressive vision of thinkers like Aldo Leopold, John Dewey, and others remain alive? Can authoritarian governance do better, and, as conditions worsen and economic and social dislocation grow, will democratic governance be undermined where it (to some extent) exists? In 1787, as he was leaving Independence Hall on the final day of the Constitutional Convention, Benjamin Franklin was asked if the delegates had created a republic or a monarchy. He replied, "A republic, if you can keep it."[2] Over the past two centuries we have kept the republic and fashioned it gradually into a representative democracy. But again today the question is pertinent. Can we sustain democracy in the ecological governance that lies ahead?

An Age of Consequences, Again

Heretofore in human history the shaping and directing of human agency has not approached (except on local scales) the boundaries set by the biophysical fact that the earth is an open system as regards energy, but virtually a closed system in regard to matter. Until recently, such boundaries did not matter and the horizons of governance were limited only by human social organization and the mobilization of collective will. Today natural boundaries do matter as much, or more, than political ones; at any rate, they should. Population, technology, and the concerted mobilization of human ingenuity and economic activity have produced a global exploitation of biophysical "resources" with historically unprecedented pace, volume, and consequence. Humankind has entered the zone of planetary boundaries and effects. That has been the journey of growth governance directed by the consumptive social contract.

Moving beyond growth governance toward a new sense of normative responsibility and political accountability consonant with the ecologically destructive power of humanity is the challenge of the future. Will we discover how to circumvent those boundaries, or will we learn how to live within them and accommodate our aspirations and our activities to them? No doubt the temptation to find technological means to overcome natural limits will be alluring; witness the incipient discourse of geo-engineering as a response to climate change, or the various innovations in extractive techniques, such as natural gas fracking or tar sands oil recovery, designed to stave off the closing of the fossil fuel era.[3]

The sociologist C. Wright Mills makes an illuminating distinction between what he called "personal troubles of milieu" and "public issues of social structure." Mills defines "troubles" as those things that "occur within the character of the individual and within the range of his immediate relations with others; they have to do with his self and with those limited areas of social life of which he is directly and personally aware," while issues "have to do with the organization of many such milieu into the institutions of an historical society as a whole."[4]

Climate change will bring about personal troubles aplenty, to be sure. But it must be understood first and foremost as a public and a structural issue—the clash between a historical form of institutionalized human activity and the natural limits imposed on human life. Social order and stability in virtually every society today, and certainly in every nation-state, rests on economic activity based on the intensive procurement and use of energy-rich fossil carbon. This is much more thermodynamically efficient than earlier fuels, and it has made possible most of modern technology and industrial civilization. We are now realizing that burning it is a fundamental threat to that very civilization.

The consumption of fossil carbon energy (coal, petroleum, natural gas) emits massive amounts of carbon and other greenhouse gases into the atmosphere, much of which will remain there for many centuries. This is causing a net gain in the planet's exchange of solar energy and it is changing the composition and behavior of earth's atmosphere and oceans. These alterations are discernible to scientific researchers and modelers—and are becoming evident to the experience of persons around the world—increasing global temperature, melting ice masses, changes in ocean currents, unusually frequent and violent storm patterns, and noticeable alterations in the conditions for land ecosystems and habitats all over the world, such as drought, species migration, and loss of biodiversity.[5] The thermal inertia of the deep ocean, the possible release of methane deposits in the permafrost, and the prospect of deep melting that destabilizes land-based ice sheets are some examples of threshold effects in biophysical systems that are nonlinear. As we come to better understand and model the behavior of complex physical and biological systems, we discover such threshold effects and other emergent properties. Human activity leading to temperature rise beyond a certain point will set in motion geophysical processes with long-delayed effects. Once begun, they cannot be stopped, contained, or reversed by human remediation, and they will not abate for decades or even centuries. We do not know precisely what those trigger-point temperatures are, but it is very likely that we are on track to reach and exceed them sometime in this century unless immediate action is taken. Substantial reductions in the amount of carbon entering the atmosphere are required via reduced emission, increased natural sequestration such as reforestation, or a combination of both. Artificial sequestration of atmospheric carbon is also theoretically possible, but would be enormously expensive.[6]

Moreover, emissions come not only from the consumption of fossil carbon energy but also from its production. Easily accessible fossil fuel deposits are becoming depleted, and more difficult and costly extraction methods are coming into widespread use—tar sands oil, mountaintop mining of coal, and hydro-fracking of natural gas in shale. In addition to the CO_2 emissions caused by the downstream burning of the fuels they produce, each of these technologies and extraction processes is a significant source of atmospheric emission in its own right, and each has other serious environmental consequences in terms of freshwater use and degradation, toxic by-products, and destruction of habitat and ecosystem services in the region of the extraction operations.[7]

In short, use of the most significant source of energy upon which humankind now relies must be curtailed very soon and replaced with energy sources that do not rely on fossil carbon. Most of the remaining fossil carbon deposits must be left in the ground. Economic and political ways must be found to prompt this massive change in human behavior, especially among people and nations that are the most intensive carbon users and are the world's richest, most powerful, and most materially comfortable. Ways must be found to offset the hardship and disruption that these economic changes will cause, especially in societies that are very highly stratified in terms of wealth and income.[8] This is a global phenomenon, so these responses must be applied not only within nations but among them. It is also an intergenerational problem. If we don't pay these prices now, others will have to pay a much larger price for the health consequences and social disruption later, likely under much less auspicious circumstances.

The health effects that will build up around the world in response to climate change of course relate to the fact that climate is intimately connected to the basics of human survival, well-being, and social order. Climate change undermines food and water sources, stable housing, biodiversity, and ecosystem services, thereby fundamentally threatening population health.[9] Social, no less than natural, determinants of health are affected. Climate change will lead to increased drought and famine, flooding, violent damaging storms, and political conflict, all of which disrupt vital public services, such as the production and distribution of goods and services, sanitation, and law enforcement. These incidents will become patterns and reinforce the discontents of fear, disrupted expectations, anger, and widespread loss of trust at a very fundamental level. Now we respond to incidents (disasters) as they occur. In the future, ecological governance will have to solve the problem posed by the patterns.

Withal, a global distribution of benefits and burdens will emerge that, if anything, will be more unjust than it is at present. A recent review of the literature on the current projections of the Intergovernmental Panel on Climate Change (IPCC) and on the health effects of climate change summarizes the situation in the following terms:

Impacts of climate change cause widespread harm to human health, with children often suffering the most. Food shortages, polluted air, contaminated or scarce supplies of water, an expanding area of vectors causing infectious diseases, and more intensely allergenic plants are among the harmful impacts. More extreme weather events cause physical and psychological harm. World health experts have concluded with "very high confidence" that climate change already contributes to the global burden of disease and premature death. IPCC projects the following trends, if global warming continues to increase, where only trends assigned very high confidence or high confidence are included: (i) increased malnutrition and consequent disorders, including those related to child growth and development, (ii) increased death, disease and injuries from heat waves, floods, storms, fires and droughts, (iii) increased cardio-respiratory morbidity and mortality associated with ground-level ozone. While IPCC also projects fewer deaths from cold, this positive effect is far outweighed by the negative ones.[10]

The increasing discussion of the health effects of global climate change calls attention to the fact that the environment is an interrelated holistic system and that health hazards come from factors that undermine the integrity or functioning of that system, such as biodiversity and ecosystemic resilience. For example, deforestation in tropical areas involves a chain of factors that ultimately affect the quality of life of persons with asthma in Central Asia; changes in the salinity, acidity, and temperature of the oceans will affect heat emergency events in Europe. A contaminated well is a localized health risk; environmental changes on the Himalayan plateau that alter the hydrology of entire river systems, on which hundreds of millions depend for their freshwater supply, represent a different challenge for public health analysis and response. The problem is global and institutional, which is to say, fundamentally political and economic. It requires more than merely specific protections and rules or laws. It requires a comprehensive engagement of governance on a number of different scales.[11]

This poses a serious anomaly to the general cognitive frameworks of human understanding of nature and a severe challenge to the assumptions and functioning of social, cultural, and political logics in contemporary technological societies. Simply put, the ideas and institutions upon which our current capability to respond collectively to climate change rests are out of step with the natural realities and threats we are discovering. Our collective capability to take climate-stabilizing action is in question. The accumulating body of scientific knowledge and evidence concerning the anthropogenic causes of climate change lends a very high degree of confidence about the challenge, but science alone seems incapable of mustering decisive political action. The human source of climate change is certain, but the human capacity to respond is not.

The ecological social contract involves finding a new consciousness and will to curb humankind's destructive economic and ecological behavior. As I shall argue in a moment, this demands civic *commonality* rather than merely self-interested *cooperation.*

The marshaled intelligence of humankind—four decades of concerted international scientific work represents precisely this—provides compelling reasons why further delay in drastically reducing atmospheric carbon (through both reducing emissions and enhancing sequestration in forests and other natural sinks) is irresponsible.[12] Further delay risks triggering long-term lag effects that are much more severe than previously recognized. Permitting global temperature to rise by 2° C by the end of the century, once considered a reasonable goal, is not an acceptable option. It appears to be still technically possible to avoid that or higher levels, but not for much longer.

To be sure, there are powerful reasons of enlightened self-interest that by their own inner logic alone should lead to the steps required to limit the damage being done not only to the climate system but also to other fundamental planetary systems of life, such as biodiversity, the planetary nitrogen load, and freshwater systems.[13] And yet look at what is happening and what seems likely to happen. Enlightened self-interest is not working. It is being thwarted by the logic of competitive advantage in markets and the politics of governance, both of which operate to prioritize short-term, shortsighted interests over long-term, synergistic considerations.

Throughout this book I have argued that three great transformations are required in our thought and action.

First, it is essential to *reorient* our predominant cultural understandings of the human place in the natural world. This is both a scientific and a philosophical undertaking.

Second, it is essential to *reconceive* the predominant economic worldview of neoliberal global capitalism.[14] This requires a new understanding of the needs and circumstances of human societies and individuals—social welfare, human flourishing, rights and liberties, growth, progress, and wealth. It also requires new institutional forms and limits on the permitted functioning and effects of economic markets, on the organization of human labor and work, and on the basic activities of extraction of natural resources and the expulsion of waste products into natural systems.

Third, it is essential to *restructure* our value priorities. This requires the widespread recognition and acceptance of the imperative of ecological responsibility, the present and intergenerational duties we have in our own individual and species flourishing, and also the duties humans have to all forms of life and to the sustainability and resilience of living systems.[15] Imposing new responsibilities on each individual and each polity to conserve the ecological and planetary systems in which they subsist may be the only way out.[16]

I want to hold out the hope that doing so need not undermine democracy or sacrifice individual freedom because enlightenment (reasoned conviction) and enchantment (a sense of higher, transcendent obligation) can be mutually reinforcing and because worldview democracy and pluralistic discursive democracy can go hand in hand. This will involve actions that support the principles of right relationship and right recognition—such as encouraging the practices of personal, civic, and institutional responsibility for sustaining the integrity and resilience of an ecological commons—and enacting shared rules and restraints based on an understanding of the good of human and natural flourishing.[17]

No one should underestimate the stakes or the difficulty of the conceptual and the practical work—the moral and the political work—ahead. Three important books on climate change ethics—*A Perfect Moral Storm* by Stephen Gardiner, *Climate Change Ethics* by Donald Brown, and *Reason in a*

Dark Time: Why the Struggle Against Climate Change Has Failed and What It Means for Our Future by Dale Jamieson—identify and discuss significant challenges to be met.[18] In the following pages I explore some of the ways that an ecological social contract might better address aspects of the moral storm of climate change. Going forward I believe that the most promising new lines of thinking cluster around the following broad questions, expanding the ecological social contract of any one society in both cultural space and generational time.

Can Global Justice Be Achieved?

Global justice is right relationship with, and right recognition of, contemporaneous humanity and nature. It is those of us in the developed parts of the world (North America, Europe, and now China and India) who have brought about—and are still continuing to bring about—the carbon emissions leading to destabilizing global warming, while those in the less developed areas are going to bear the brunt of the dislocations. The distribution of these benefits and burdens associated with climate change will be disproportionate, and this injustice piles on top of the long-standing injustice of the distribution of global wealth and income and of health and welfare. The old paradigm of development economics—growth through the dissemination of carbon-intensive energy use and technology—won't work. That rising tide will swamp all boats. Can we find a way to share wealth and power more equitably in a world of lower growth?

Can Intergenerational Justice Be Achieved?

Intergenerational justice is right relationship with, and right recognition of, future humanity and nature. As difficult as the challenge of practically meeting the requirements of contemporaneous global justice may be, the problem of intergenerational justice is even more perplexing. When we are talking about contemporaneous persons, the shaping of their quality of life, options, and choices is clearly a matter of justice and human rights. What moral difference does their status as future beings make exactly? Moral philosophy today is not clear on how best to answer that question. The task of getting the rich to recognize the rights and common humanity of the poor is common to both problems of justice, but it is complicated in intergenerational justice by the issue of the moral standing of persons who only exist statistically and probabilistically, not individually and concretely.

If we can somehow forge a new global social contract of ecological trusteeship in place of the current contract of self-interested consumption, can we find a place within the new contract for those yet unborn? The metaphor of the social contract captures reciprocal relationality and interdependence among contemporaneous persons. But when we talk about relationships with persons that do not yet exist, inhabiting ecosystems and states of the world that do not yet exist—and may never exist depending on what we do in our lifetimes—what is the moral force of those relationships with those not-yet persons?

Surely it is incorrect to say that there is no conceivable relationship here or that such a notion violates the meaning of the concept of relationship. What we do now will in fact affect "not-yet" persons and the natural world they are born into and inhabit. Granted, this cannot be reciprocal since the future party cannot affect us, except through the medium of moral imagination and conscience. And yet our actions in the present do have the power to shape the quality of life and the options of future people and the integrity and resiliency of the future ecosystems they inhabit substantially. Climate change brings the pluperfect tense of ethics to the fore in dramatic fashion.

Every society needs to have a discourse to give expression to its sense of what history asks of it, a discourse with which to affirm and to contest power, equality, individual and group identity, knowledge, duty, and trust. Indeed, societies ideally have not just one such discourse, but several layered and overlapping ones. Repressive and stagnant societies tend to flatten and winnow this discursive landscape; more dynamic and open societies tend toward more diversity and argumentative conflict. And every society needs a discourse to articulate the appropriate role and place of humans in nature: Are we creators or creatures, are we destined to overcome limits or to accommodate ourselves to them? How should we use nature and what does nature ask of us? And how should we engage with our own humanity and what does our humanness ask of us and require? Finally, what is the calling of this moment in the ecological history of life on earth and in the history of humankind? What have we the power to do? What have we the responsibility to do?

How Do We Know What We Owe One Another?

How do we get people to see their obligations? How do we motivate them to act on those obligations even when it involves some denial or sacrifice of one's own wants and interests? One of the reasons why appeals to the prudent protection of enlightened self-interest have not succeeded in motivating political support for equalizing and redistributive policies is that well-off individuals can see the reality of relative inequality all around them—in the form of poverty, crime, inadequate education, health disparities, and so on—but they do not perceive that this inequality undermines their own quality of life or future prospects. Thus instead of feeling empathy and solidarity for the least well off, they feel threatened by and antipathetic toward them. Their main preoccupation is keeping their footing on the rung they have managed to attain and not slipping down the social ladder. In a discussion of health disparities and the social determinants of health, David Runciman observes, "the politics is considerably harder here: you can't simply say that inequality means we are all suffering together. Instead, it may mean that the poor are doing so badly that the rich aren't interested in looking at the wider picture. They are focused on making sure they don't wind up poor."[19] Thus far this same syndrome has undermined political support for policies to reduce carbon emissions, such as a carbon tax or any other measures that would threaten to raise consumer costs or increase unemployment.

If we are to use self-interest as the primary motivating factor in garnering democratic political support for climate-smart public policies and the effective regulation of commercial and private behavior, then we need to break out of this syndrome of social antipathy and competition. Simply striving for conditions to facilitate long-term self-interest over short-term self-interest is not sufficient. The politics of falling down and falling behind in a stratified society is not so much a question of the time scale of the personal and social cost-benefit equation. It is primarily a failure to see the connections between one's own social-economic situation and that of others, a failure to perceive the underlying forces of economic and social power that are working on everyone in the society, albeit with differential effects.

How Do We Break Free of This Conundrum?

I do not believe that we can simply try to suppress self-interest in the motivational structure of individuals and replace it with some overriding moral ideal of duty or principles of justice and beneficence. A sentinel task for our public discourse and moral learning is to temper and reconstitute self-interest by interpreting it in new ways, much as we need to reinterpret freedom. This concerns reconceptualizing the constitutive features of self-interest (or happiness) by not only expanding its horizons of time and place, but also by reconceiving the subject or self whose interests are at stake. Doing so provides a vocabulary with which to speak about who one is and what one is doing in new ways. And this leads to speaking about who *we are* and what *we are doing* in new ways as well. It gives us a lens through which to see ourselves, our situation, and our possibilities in a new light. If the current failure of self-interested motivation is the failure to see connections, and hence the failure to see and care about the consequences of how our activities are institutionalized and structured, then the remedy can come in the form of an enrichment of our connection-making moral imagination.

Reductions in GHG emissions will come about only through change at both the level of individual behavior and of social norms and institutions. In practice this means that public policies must have recourse to values and purposes that the members of these societies will understand if they think and act like interdependent and relational selves. Discursively, the task is to shape this relational self-identity and foster a moral imagination that can see autonomy, respect, rights, and responsibilities in relational terms. If it was not morally evident before to all reasonable people (as the current global economy guided by neoliberal free market ideology suggests that it has not been), at a time of climate change it surely must be recognized now that there is no immunity, no safe harbor, no fortress of privilege and security. The health and well-being—as well as the possibility of a life of autonomous self-direction—of everyone is inextricably linked to the flourishing of others, the flourishing of emplaced communities, and the flourishing of the natural world.

The most daunting challenge of climate change is not technological or even economic, it is political and moral. Prosperity without unsustainable economic growth can be attained through rational orchestration of measures to slow down and reduce the consumption of fossil carbon and through technological transitions to new sources of low-carbon emission and renewable energy.[20] But these measures, such as carbon taxation, to reduce—or better, recast—economic growth must be accompanied by robust social policies that forthrightly, not begrudgingly, embrace goals of social justice and equality, education, meaningful employment, and democratic citizenship. If that is to happen, a certain moral maturity—the moral maturity of which Pope Francis speaks—must be achieved in many of the political cultures in the world today.

12

An Inquiry into the Democratic Prospect

Looking back on the environmental debates of the 1970s, I am struck by an overwhelming sense of opportunity missed then and precious time squandered since. Our political and moral efforts to come to grips with the limits to growth were feeble and inconclusive.[1] And how rapidly in the aftermath of the OPEC oil embargo and falling rates of global corporate profits (which should have been seen as writing on the wall) the prevailing intellectual winds shifted and returned public attention back to the more familiar economic issues. In 1979 President Carter dared to speak some inconvenient truths (*avant la lettre*), and Ronald Reagan made him pay the price.[2] President Reagan was elected very clearly on the basis of the social contract of consumption. On environmental matters, presidential courage and leadership in the United States never recovered. The normal agenda of the growth-oriented political economy of liberalism was reestablished with remarkable ease and virtually without any commentary on the implications that this extraordinary failure of political nerve was going to have for the future.

Arguably the most important debate of the late twentieth century was not so much decided as terminated, and elsewhere in the political economy a shift from egalitarian liberalism to market liberalism took place with remarkably little real opposition. Instead of rising to the occasion of this historic challenge posed by the limits to growth, liberal governance just shrugged its shoulders and quickly reverted to the conventional wisdom that "a rising tide lifts all boats," and sought the contentment of affluence through economic growth. In virtually all developed countries and with the blessing of virtually all major political parties, economic policy has concentrated on fostering high technology-oriented productivity and attempts to stimulate aggregate demand for products. Environmental regulations and protections have been little more than side constraints, acceptable so long as they did not interfere with profits, employment, or the voracious engine of consumer demand.

Governance in a Time of Neoliberalism

During this period, little by little, Keynesian liberalism has morphed into neoliberalism and free market orthodoxy.[3] Today, not only environmental policy but even the protection and management of the post–World War II welfare state seem to be lost causes. Income has been stagnant in real terms for the majority of the workforce in the developed world, while the ecologically harmful volume and pace of economic activity have increased tremendously.[4] Inequality in the domestic distribution of wealth and income has reached levels in many countries not seen since the heyday of industrial capitalism in the late nineteenth and early twentieth century. Social mobility is growing more rigid; in much of the developed world, including the United States, the single best predictor of a young person's lifetime income prospects is the income of his or her father.[5] Only a few countries, the most egalitarian ones, have not yet succumbed to that trend. If one wants to live out what is called by those in the United States "the American Dream," one needs to move to Denmark.

In addition, new developments have arisen to complicate governance enormously, such as the global mobility of capital and investment, which undermines the relative power of the nation-state as a meaningful policy maker and as a locus of economic leverage. Meanwhile, regional and global ecological problems have gotten much worse than they were in the 1970s—climate change, biodiversity loss, freshwater shortages, damage to the ocean ecosystems. Hence, the continuing viability of the liberal tradition—a proud and hard-won intellectual orientation promoting freedom, equality, and human rights for three hundred years beginning in the seventeenth century—is in serious question. Can we be sanguine about the possibility of genuinely coping with limits to growth while still remaining committed to these basic values, institutions, and practices?

Two points, however, do seem reasonably certain. First, while we do not know what form the transition to a new structure of governance will take, we do know that some such transition will be necessary and inevitable.

The second clear starting point is that whatever type and form of governance emerges, it will require normative legitimation to be sustained. Before moving further in the discussion, let me clarify what I mean by governance. Governance is not the same thing as government. Governance is the overall process of coordinating, shaping, and directing individual and collective agency. Governance is inherently normative and value-laden. It sets parameters around the means and forms of human agency, excluding some practices (such as genocide, murder, torture, slavery, rape, bigotry, and racism) from the sphere of social life as intrinsically illicit. Governance also defines the telos, the ends, of collective agency; it stipulates worthy ideals, places parameters around the objectives to be intended and sought, and excludes some types of objectives as wasteful or unworthy. Finally, governance embodies the character of the collectivity, representing the kind of society an association of people aspires to be or become. Governance both rests upon, and enacts anew, the understanding of solidarity that holds individuals together in shared meaning and common purpose and mutual endeavor. Governance is an enabling act of mind that creates communities; its work is the construction of institutionalized normative practice and symbolic orders of meaning.

So conceived, governance is a process that involves many institutions—in the economy, civil society, and religious and cultural organizations—in addition to the government legally defined. Questions about the form that governance in an ecological political economy should take are therefore not limited to structural questions about the location of authority, the distribution and interaction of powers, the selection of individuals to fulfill specialized roles, or the enactment and enforcement of common rules, as vital as these matters are. Glancing toward Montesquieu, I would say that governance is not only about the letter of the laws, but also about their spirit; not about the body of law, but about its mind.

Beginning in the 1970s, a number of social theorists began to maintain that ecological constraints will create a legitimation crisis for liberal democracy and that either a nondemocratic authoritarian state or at least a democratic regime with new nondemocratic power centers will emerge from that crisis. A future authoritarianism does not necessarily entail a military dictatorship or police state. Coercion alone, even if ethically justified, cannot sustain behavioral compliance across a large population and govern complex networks of economic activity under modern social conditions for a sustained period of time. Popular conviction and voluntary consent, not coercion, are the key to modern governance, certainly on the national level, let alone on larger scales than that. Hence whatever effective form of governance emerges in a future ecological society, a new form of social contract will be needed as its foundation: a transformation within the political culture that will produce voluntary consent to the new forms of governance and to a new reach of political authority.

Such conviction is brought about in one of two ways: by purchase or by persuasion; by deploying financial incentives and self-interested motivations, or by manipulating ideas, ideals, and arguments. If the growth of material consumption and affluence will not be the currency with which to buy the necessary compliance, then what form of persuasion can secure them?

Compliance or obedience, especially in times of crisis or upheaval, may begin in fear, but ultimately a different kind of motivation will be necessary for governance to succeed, particularly in a democratic and open society. That motivation, as I have discussed, involves both reasoned conviction and felt contentment, and both of these are at bottom, functions of self-identity, meaning, and cultural worldview. Over a longer term, these motivational and cultural foundations can be well established. The problem is the period of transition from one social contract and political economy to another. The emerging ecological social contract must provide the transitional generation of citizens with a way to respond positively to the foreclosing of an unsustainable growth future they have been taught to expect. And it must teach future generations different lessons and nurture different kinds of expectations.

On the policy side, a new social contract for ecological governance will need to navigate conflicting social and economic interests and blocking coalitions on many fronts. This is not a requirement peculiar to democratic governance, for an authoritarian regime will have its civil society to contend with also. There will be internal bargaining and positional politics in the bureaucracy of an authoritarian regime, and authoritarian bureaucracies are as adept as any electoral parliaments at generating blocking coalitions and "friction" to the point of slow-motion action and endlessly deferred outcome.

So democratic or not, the challenges of the new governance will be formidable. For example, in economic policy, pricing mechanisms will have to be restructured to reflect the true ecological costs on many fronts: the true cost of extracting scarce and non-replenishing raw materials; the costs of preserving the capacity of ecosystems to contribute value through naturally occurring processes; and the costs of mitigating the damage done to both biotic and human communities by the waste products of anthropogenic economic activity. Governance will not be conceptualized, as it is today in the ideological viewpoint of neoliberalism, as separate and distinct from the market. Working symbiotically, governmental regulation and market incentives must channel public and private investment into new, more efficient technologies and production processes, alternative energy sources, and the like.

Moreover, the key to success in this new realm of policy and governance is to find some way to deal with the fact that, even in an ecological society, any government's fiscal wherewithal (its ability to generate tax revenue) and its political stability will depend on a private sector that generates productivity and provides employment income for the vast majority of the population.[6] (Even in the welfare states of the aging advanced societies, where the ratio of the employed to the unemployed will be quite unfavorable later in the twenty-first century, social support is dependent on revenues generated by investment in private sector securities markets.) The question is this: What are the right forms of generating private capital and public revenue that will support the state and the nonprofit sector without ecologically unacceptable modes of economic activity? The answer seems to lie in what some economists call "decoupling," a situation in which productivity, which supports profit and the revenue upon which the state depends, increases but does so without concomitant increases in extractive and excretory activity that is ecologically out of bounds.

A parallel question concerns the relationship between the private and the public sector in ecological governance. Starting from our current institutional formation of state and market, can we evolve into something more closely resembling the classical notions of polity and household (*polis* and *oikos*)? Many tend to think of the market as an autonomously operating, impersonal "system" or structure that needs no intentional agency of governance and no deliberately governing subjects. That is a fiction of conceptual and mathematical modeling and libertarian rhetoric. (In actually existing capitalism, there is very significant and deliberate governance exercised in both the public and the private or corporate sectors.) Be that as it may, the issue at the moment involves what does look like a systemic or structural aspect of historical capitalism, namely, its cyclical pattern of expansion and recession, boom and bust.

The Transformation of Capitalism?

The concept of degrowth seems to require, if not the abandonment of capitalism, then at least a transformation of capitalism so as to escape that cycle into something more closely resembling a steady-state system that fluctuates only within planetary boundaries and ecosystemic tolerances over time. Hence the dialectical relationship between impersonal system and intentional agency—structure and subject—will be central to political theory, constitutional formation, and law in the decades ahead. This poses yet another intellectual challenge for us since we are not very well equipped to theorize that relationship adequately. Aside from Marx, few in classical political economics devoted careful attention to it, and in the modern transition to marginalism in the work of Jevons, Marshall, and Pigou economic theorizing moved away from this set of questions decisively.[7]

Today, however, we face the prospect that a new transnational structure of political economy accommodating limits and operating on something like closed-system, steady-state mechanisms will have to be brought about at least partially through deliberate agency and political decisions. The alternative is to see a steady-state system as emerging spontaneously as an evolutionary stage or an outcome of the logical systemic dynamic of capitalism, which seems to have been John Stuart Mill's progressive hope.[8] But Mill, like nineteenth-century liberal thought generally, tended to conflate the systemic logic of historical process with the moral agency of individuals, thereby rendering the problem of the relationship between them conceptually opaque. Future governance and policy in an ecological society may be able to float on the underlying dynamics of a steady-state economic system once it is institutionalized, but the transition from the system dominated by the logic of growth to the steady-state logic will have to be forged through political action and normative will.

Let me flesh out this last point. In order to systematically understand historical change, it is useful to differentiate between a transitional period from current to new, on the one hand, and a post-transitional period, on the other. This distinction is also pertinent when discussing governance options. This is not to say that the eventual post-transition society is going to be static or outside of history. It is merely that the institutional formation of governance necessary in the transitional period and the governance formation of the post-growth political economy need not be the same. Both may be democratic, or neither may be. And the scenario of a relatively more authoritarian, less democratic transition that leads to—and gives way to—a more democratic governance eventually is also a possible and a plausible story to consider. It is attractive for democrats currently in despair, for it has a desirable long-term prognosis. But this two-step scenario—sacrifice important political values temporarily out of necessity, only to realize them more authentically and fully than ever before later—can be seductive and is not without problems and blind spots of its own, as we shall see.

Democratic Prospects Beyond Liberal Democracy

What then are the prospects for democratic governance in an ecological political economy? Answering such a question adequately is of course out of the question in a few pages, but in what follows I aim to assist what must be an ongoing reflection by laying out a kind of road map or anatomy of ecological governance. This anatomy consists of three heuristic types of governance orientation, which I shall refer to as *ecological authoritarianism*, *ecological discursive democracy*, and *ecological constitutionalism*. Although the work of many theorists could be mentioned, I take as representative theorists of these three orientations William Ophuls, John A. Dryzek, and Klaus Bosselmann and J. Ronald Engel, respectively.[9] I give each of these types of governance the epithet "ecological" because I am interested only in theories of governance that take the reality of ecological limits to growth as their starting point and adopt what might be thought of as an ecocentric normative and epistemological stance (as opposed to an anthropocentric one).

This stipulation obviously puts brackets around many of the types of governance studied by mainstream political theorists and political scientists. Discursive democracy is not usually taken to be a practical form of democracy at all. But I think we must take it seriously as an option for an ecological society. Its democratic rival, variously called liberal democracy, interest group democracy, or representative democracy, is considered to be the only viable form of democracy in the modern world. It is certainly the most widespread form of democratic governance in operation, but it does not have, in my view, an ecological variant as such and therefore will not figure in my schema of types of ecological governance. For reasons alluded to above and discussed in more detail in a moment, interest group democracy is not succeeding and is structurally incapable of presiding over the transition from a consumptive growth-oriented political economy to an ecological political economy. Within the authoritarian and constitutional types, we must single out an ecologically oriented variant as well, since pro-growth technocratic variants of these formations certainly do exist and even predominate in the world today. Technocratic elitism is not the kind of authoritarianism—and a growth-oriented technocratic corporatism is not the kind of constitutionalism—envisioned here. They are not part of the ecological solution; they are part of the problem.

Before turning to the three types of ecological governance, consider further the contrasting benchmark of interest group democracy. All of the ecological types of governance I identify have one thing in common, namely, their critique and rejection of interest group democracy. Interest group democracy is concerned with aggregation and accommodation of interests among individuals and groups in societies where religious differences, ideological diversity, social competition, and conflict are widespread. This is the political system of the Western world, certainly in the bicameral legislature and separately elected head of government system of the United States, but also in parliamentary systems, systems with proportional legislative representation rather than single-member districts, and so on. Interest group democracy is responsive to individual interests, concatenated or organized by the formation of various group structures that compete for the attention of popularly elected officials. Their competition in this regard consists both of the marketplace of ideas and the marketplace of campaign contributions, and other financial incentives for public officials. Unlike discursive democracy, in which the citizen role is actively and extensively participatory at multiple levels, in interest group democracy citizenship consists essentially in the right to vote, with a relatively small number becoming directly involved financially or personally in the process of electoral competition. Candidates and parties vie there for the support of self-interested voters, which is increasingly determined by media advertisements and exposure.[10]

Interest group democracy is a kind of negative system of governance. It is set up to form compromise among conflicting interests so that no one group bears the cost of policy. This makes a win-win type of growth scenario very attractive and deters policy makers from setting clear priorities, making trade-offs, especially sharp ones, which have been called "tragic choices," such as rationing and redistributing resources (wealth and power) explicitly.[11] It has multiple veto points in its governing process that ensure these features. It is prone to incrementalism and bias in favor of preserving the status quo.

Varieties of Ecological Governance

Against this backdrop, I now turn to the three modes of governance that I think are reasonable options for an ecological transition and the resulting new political economy.

Ecological Authoritarianism

Ecological authoritarians maintain that the successful governance in an ecological era will require centralized, elitist, and technocratic management at least in the areas of economic and environmental policy.[12] Mindful of the internal contradictions interest group democracy faces as it attempts to cope with problems of productivity, capital accumulation, and growth, ecological authoritarians stress the need for policy makers and planners to be insulated from democratic pressures and granted an increasing measure of autonomous authority if they are to steer the economy on an ecologically rational and efficient course. Ecological authoritarians are impressed, perhaps overly so, by the popular demand in pluralistic democratic systems for democratic rights and material affluence. They speak of democratic overload in reference to those pressures and demands: democratic overload of policy makers leads to economic overload or overshoot of the carrying capacity of ecosystems. The former has to be broken free from in order to prevent the latter.

Indeed, ecological authoritarians see a vicious cycle, a destructive feedback loop in this. As pluralistic democracies succeed in their aim to increase economic prosperity for the population, the democratic assertiveness of citizens for more growth and prosperity also increases. As the economic management of ever-higher levels of affluence becomes more complex, the tension between democratic politics and "scientific" planning comes to a crisis point.

The ecological authoritarians here make an important point. The fact that pluralistic democracy has demonstrated its inability to perform ecologically precautionary governance in a consistent or timely way is not fortuitous; it is built into the deep structure and political logic of this type of system as such. If pluralistic democratic governments follow the dictates of ecological science and planning, they will restrict growth in ways that risk losing their popular base of support. If, conversely, such governments attempt to maintain their legitimacy by bowing to short-term democratic pressures, they will not be able to take (and require the private sector to take) the steps necessary to protect the environment. Eventually economic downturn, inequality, and hardship will result from ecological degradation, and again the governments will lose their popular support and legitimacy.[13] Note, however, that the political costs of the first prong of this dilemma are more immediate than those from the second prong, so pragmatism in a pluralistic democracy counsels the first course of action. Such pragmatism is ecologically insane.

The work of William Ophuls, a political theorist who has focused on ecological issues and the limits to growth in a serious and sustained way, is significant in this regard. In his devastating critique of American environmental policy, Ophuls lays out the logical contradiction of interest group democracy in detail.[14] Ophuls is not an ideological authoritarian as a matter of political philosophy or principle. He is driven to it by an embrace of what Hans Jonas called "the imperative of responsibility," and by the force of both eco- and political-systems logic. Alluringly, Ophuls holds out the possibility of a highly decentralized, communitarian society for the distant future, one in which individuals would enjoy what he calls "micro-freedoms" within a framework of "macro-constraints." However, during the transition era, and even in the eventual steady-state era, he is quite clear about who will decide what these "macro-constraints" will be and who will impose them: "The ecologically complex steady-state society may . . . require, if not a class of ecological guardians, then at least a class of ecological mandarins who possess the esoteric knowledge needed to run it well."[15]

Ecological Discursive Democracy

The second alternative form of governance involves the institutionalization and empowerment of participatory and discursive governance within a diverse and pluralistic society and culture, a "panarchy" as the Resilience Alliance scientists call it.[16] This is the kind of democratic governance that grows directly out of concerted action with others, shaped by debate and deliberation.[17]

Ecological discursive democracy differs from interest group democracy in fundamental respects. It challenges the primacy of a materialistic and consumptive notion of interests as the basis of the psychological and moral dimension of a democratic polity. It also argues for both the feasibility of—and the normative justification of—the maximum feasible direct participation and direct engagement in governance issues, at least in making fundamental value choices and setting broad priorities, if not in all the technical details of regulation and enforcement. Its conception is that democratic citizens should act, not as passive electoral consumers, but as dialogic partners in an attempt to discern the common good and the responsibility of the democratic community for the integrity of the biotic community of which it is a part.[18]

Ecological discursive democracy is committed to the strategy of creating counter-publics in order to bring about change and to challenge the hegemony of mainstream culture and politics. In a transition, it will exist side by side with interest group democracy, and eventually could very well coexist with ecological constitutional democracy. Discursive democracy shares strategy and tactics with mainstream politics and mixes in real-world political activism and large-scale protest movements.[19] It does not have to rely on direct action or small-scale participatory decision making alone; it can work with layered systems of representation from local to national and even to global scales. It can find expression not only in electoral mechanisms but in the work of civil society and nongovernmental organizations as well. Ecological discursive democracy does not have to supplant mainstream politics, it just needs to set cultural and social forces in motion that will alter perceptions and change the parameters of what is considered realistic in elite policy circles.

Ecological Constitutionalism

This type of governance involves building new ecologically oriented norms and values into the constitutional structure and/or the legitimating consensus of a democratic mode of governance.[20] Ecological constitutionalism may continue to resemble the representative form of interest group democracy, perhaps with a rather more robust discursive and participatory element at the local levels and in the periodic electoral process. The institutional structure of ecological constitutionalist governance would involve the creation of several elite governing entities that can check and balance the more representative institutions such as the presidency or legislature. These elite entities would be insulated in various constitutionally sanctioned ways from the pressures of interest groups that do not benefit in the short term from policies and decisions consonant with the limits of growth and the protection of ecosystem resilience.

The most prominent example of such a democratically insulated governance structure is the judiciary. Environmental courts are being created and given substantial governance authority in various parts of the world today.[21] Another example is the creation of relatively autonomous regulatory agencies and commissions that have the legal authority to determine policy in some areas independently of the executive or legislative branch. Another important example is an independent and scientifically based professional civil service, selected in accordance with a meritocratic system and protected from politically motivated interference or dismissal.[22] I think some of the best examples of what I have in mind by ecological constitutionalism exist at the international level and in the domain of environmental "soft law," such as the Earth Charter process and movement.

Constitutional and institutional inventions such as these are the historical legacy of past times in which the operations of free markets or interest group democratic politics were seen to be detrimental to the public interest, and a more professional and expert type of elitism was turned to for a powerful role in governance. But the spirit of ecological constitutionalism is not deference to expertise or meritorious authority per se. At bottom it rests on a normative and psychologically powerful commitment to the rule of law and the exercise of power that is governed by norms of judicial interpretation or scientific verification. These norms are still in need of interpretation, but they are generally clearer and more settled than the ideological values and self-regarding interests that hold sway in the interest group competitive political system. Like ecological discursive democracy, ecological constitutionalism looks to a form of governance based ultimately on persuasion, consensus, and decisions that are viewed as legitimate by the governed. Ecological authoritarianism rests more straightforwardly on respect for authority, trust, prudence, and, sometimes, fear of penalty for failure to obey. If discursive democracy places its hopes on the communication process of fair, impartial, and respectful give-and-take of ideas, ideals, and reasons, ecological constitutionalism relies on the cultural formation of an animating worldview and sense of attachment to place.[23]

Ultimately ecological constitutionalism relies on a transformative process of moral and political imagination at the pre-political cultural level that ecological discursive democracy hopes to achieve through democratic political life itself. That is a transformation of our "soul" as a political community, turning us from being a people of competitive consumption into a people of sustainable ecological responsibility.

Dilemmas of Ecological Governance

So far I have sketched these three modes of governance in a way that brings out their respective strengths and insight. I turn now to some critical reflections on their characteristic problems and weaknesses.

Ecological authoritarians are exceedingly pessimistic about the current state of ordinary moral sensibility and political judgment in Western society. They rest their assessment and theory of governance on a belief in the widespread cultural dominance of excessive individualism and materialism, and seem to think that consumerism has thoroughly triumphed over every competing cultural or ethical value system. As an account of the logical tendencies at work in a particular form of economic system and society, this argument has validity, but not as a complete empirical description of currently existing culture and politics. It underestimates the remaining moral capital in a growth-oriented society that can be channeled in future ecological directions, thereby making a different form of democracy compatible with ecological requirements. Many still quite powerfully held values compete with individualism and materialism in the attitudes and motivations of contemporary citizens. Furthermore, it is far from clear that massive numbers of people will withdraw their allegiance from democratic governments even if they fail to produce high levels of economic prosperity. Indeed, a democratic system may well be able to survive even a period of considerable austerity, if the need for such measures were clear and if the burdens and hardships were shared in a truly just and equitable way. These are large ifs, as the current situation in the Eurozone and the management of the fiscal crisis in Greece demonstrate.[24] The relationship between economics and political legitimation is complex.

As a matter of intellectual history, this equating of all forms of democracy with materialism and a narrowly self-interested individualism is overly selective. Even in the modern period, democratic governance has rested on more than simply the ideology of growth and expansive capitalism. While possessive individualism has certainly been one important strand within Western democratic societies, it has not been the only strand, and it has not gone unchallenged from within democratic thought itself.[25] Thus it is misleading to argue that democratic governance cannot honestly appeal to any sense of public purpose larger than throwaway consumption. Believing that democratic governments could neither call for nor obtain popular support except on the basis of a promise of more economic growth, ecological authoritarianism turns to non-democratic governance out of a sense of despair. But, so far at least, neither the sociological nor the historical arguments of this theory are sufficient to warrant that conclusion.

The dilemmas of discursive democracy are rather different. It is designed to thrive on pluralism of belief and difference of opinion. But it must inculcate at least a minimal set of value commitments to the procedures of debate and deliberation. Realism, reason, and integrity-preserving compromise are its creed. Toleration and diversity are its lifeblood. Discursive democratic governance has been shown to function well on a medium to small scale, in population units of 100,000 or less, and when its political relationships are closely embedded in nonpolitical social or civic relationships within the community.[26] Under these conditions, it is alert to natural, social, and historical place. It can be attentive to ecological resilience and social justice at the same time.

On the other hand, discursive democracy is extremely vulnerable to forces that disrupt the fabric of communities, are socially divisive, undermine trust, and drive people to close ranks into postures of defensive resentment. The global and domestic economic dislocations of the past twenty years, the sharply rising inequality in the distribution of wealth and income, the churning of the job market, and the marginalization of those without marketable skills are some of the many factors that tatter civic society, privatize self-consciousness, and undermine the possibility of the kind of citizenship that discursive democracy in its proper form and function requires. Are these factors temporary aberrations, or are they becoming the normal institutionalized patterns of global capitalism? If they are, then we have a perfect democratic storm: humanity is exceeding the safe operating margins of planetary systems at precisely the historical moment when the political economy of the world makes it least likely that democratic governance, especially discursive democratic governance, will be able to respond.

I turn finally to problems with ecological constitutionalism. It is a form of democracy that strives to express and to energize deep cultural and ethical convictions. It offers not just a new direction of governance but a new form of life, a new understanding of human well-being, and a new story concerning nature, its laws, and its meanings. As we have seen, appealing to strong ontology in this way is playing with fire politically. If the leadership of ecological constitutionalism—or ecological authoritarianism, for that matter—succeeds in mobilizing large numbers of people with this strong vision, then a form of progressive ecological democratic change can bubble up from below and will be managed by structural constitutional checks on ill-advised democratic decisions to fall back on if need be. And this deep cultural foundation of worldview and values will restrain its own radical interpretations and populist or anarchic tendencies via the appeal of patriotic allegiance to the norms enshrined in the ecological constitution (some version of the Earth Charter) itself.[27]

If this cultural transformation of hearts and minds does not proceed well, however, leaders will be tempted to assume interpretative and expressive authority for themselves. They will become the guardians of the truths and values of the worldview and the agents of its enactment in the world. Democratic citizenship will become an unnecessary step in the process. The temptation to become a transformational leader/prophet in this more authoritarian sense is particularly strong when circumstances make one pessimistic about the willingness or the capacity of the masses to internalize new values and support change. Ecological constitutionalism may be theoretically (if not sociologically and historically) unstable. If it works, then so can a spreading kind of ecological discursive participation. If it falters, it can slip into ecological monism. It is prone to centrifugal forces from both its left and its right.

Toward a Democracy as Yet Untried

At the moment I conclude that the form of governance best suited to an ecological society will be one that is democratic and deliberative in its substance but not necessarily completely direct or representational in an electoral sense in all of its institutional forms. This takes from ecological discursive democracy the key idea that reasoning and the orientation of individual citizens and public officials must involve an attempt to discern the common good. Doing so ultimately rests on their capacity to see the connection between the common good and their individual good. Because the social and cultural forces blinding one to that connection are too numerous and powerful, some measure of insulated power by carefully selected officials will be necessary, *faute de mieux*. This departs from the activity of popular rule, but it does not depart from the requirement of deliberation, which actually cuts across popular and elite functions in a system of governance. In this way ecological discursive democracy can be linked to ecological constitutionalism. There is more then to constitutionalism than deontological rights and rules that are prior to a calculus of utility and that protect individuals and minorities from the control and coercion of popular majorities, as fundamentally important as these principles and protections are. Ecological constitutionalism must also be discursive, and among its tenets (again the Earth Charter provisions provide an instructive example) must be provisions to empower and promote the social conditions and capabilities necessary to make active participation and citizenship possible and meaningful to all.

I want to close with some remarks on three large issues that stand in the background of any analysis of the most appropriate governance for an ecological political economy. These are scale and subsidiarity; unity and plurality; and the recovery of the political.

Scale and Subsidiarity

In any discussion of the pros and cons of democratic governance, the issue of scale has loomed large for centuries.[28] Deliberative, participatory democracy requires small-scale governance; its legitimating processes and imagined communities cannot extend beyond a certain territory. Abstraction, facelessness, and impartiality erode the motivation and the type of judgment and agency that discursive democracy requires of its citizens. The nation-state and representative democracy tempered by a legal system protective of civil rights and liberties have been the historical solution to the problem of democratic scale. Yet as we have seen, both the political culture of interest group consumerism and the political economy of capitalist globalization have undercut that institutional form, at least as a governance strategy to meet the terms of an ecological social contract. Ecological constitutionalism with a discursive commitment and more participatory educative and advisory activities at more local levels of governance (subsidiarity) seems the most promising way out of this dilemma.

Unity and Plurality

A more serious and intractable problem for ecological governance is the long-standing issue of unity versus plurality.[29]

This, to me, is a more fundamental problem than representation versus participation within democracy, or even than democracy versus some kind of nondemocratic authoritarianism. In environmental ethics and green political theory, there has been considerable discussion of whether the substantive ecocentric values of environmentalism should take precedence over the procedural values of democracy. Are we still committed to democratic governance if its outcome is wrong from an eco-ethical point of view?[30] But today the issue is made even more pointed because the question is not only one of substantive environmental values, but of ecological facts. Can we still defend democracy when it produces policies and priorities that are out of step with biophysical reality? In the United States today, the political debate on this question has not yet risen to a sufficient level of intelligence to qualify as deliberation.[31] It remains deadlocked in what we might call "abuse" (personal attacks and ad hominem reasoning) and frozen assertion and counter-assertion (No it doesn't. Yes it does.) that can be called "argument" or deliberation only in a Monty Python sense of the term.[32]

Going back much further, critics of democracy, beginning with Plato, have based their criticism not only on size and scale considerations but on the observation that democracies lack the necessary convergence on a unitary understanding and set of values. If there is a single form of the good, democracies are precisely the types of states where that unifying vision will not be made accessible to the people, and it may even direct violence against those few who do grasp it. However, even within democratic theory itself, a desire for a unitary conception has also emerged historically, most notably in Rousseau's notion of the general will.

Democratic theory for the most part has taken the opposing philosophical and metaphysical stand and has embraced and celebrated human plurality. When Ophuls evokes the legacy of the antidemocratic philosophers with a unitary vision, such as Plato, he gives the argument a new and very disturbing twist. For Plato, the cost of pluralism was ignorance, deception, and the failure to attain the true blessedness that comes from a vision of the Ideas in their actual being. For Ophuls, the stakes are higher at least materialistically—the prospect of a greatly degraded condition of life, if not the extinguishing of most life, on earth. One need look no further than to the well-orchestrated process of climate change denial to take his point.[33] To respond to his Neoplatonic challenge, we need first to recognize that the main forms of plurality invented in the modern world, the liberal state and the capitalistic market, have now become unsustainable. Then we need to seek out new forms of plurality that are sustainable and can be genuinely celebrated for their open texture, their sense of solidarity and community, and their expressive freedom and traditional meanings and excellences.

Recovery of the Political

In our search for a form of plurality that is more in keeping with ecological governance, I think we need to begin by noting that our current language for talking about politics is deeply impoverished. Near the heart of the ecological social contract, it seems to me, is an understanding of the political in the following fundamental sense: the problem of politics is not to seek power, but to resist it; and not to deny vulnerability and dependency, but to embrace them creatively. The problem of politics is to resist the kind of power and domination that actually render its agents impotent and enthralled. The problem of politics is to accept restraints on behalf of communal agency and relational freedom. The political dimension and potentiality of our species being as *zöon politikon* says no to pride, anthropocentric narcissism, and desire; and says yes to the accommodation of natural limits in ways that are just and promote the beauty, health, and integrity of both the political and the biotic communities.

In this formulation, I deliberately invoke the land ethic and appropriate it in the cause of what I am here calling discursive democratic politics as the solution to the great challenge of our time. Leopold wanted human beings to think of themselves as, and to act as though they were, "plain citizens of the biotic community." He wrote: "a land ethic changes the role of *Homo sapiens* from conqueror of the land-community to plain member and citizen of it. It implies respect for his fellow-members, and also respect for the community as such."[34] He used this democratic trope, with its classical pedigree and protestant resonance, to good effect. To be a citizen in fact is to be a "plain" citizen, an equal member: one person, one vote; parity of voice and participation. And to be a citizen is to be just. Ecological governance—whether authoritarian, discursive, or constitutionalist—calls for just, plain citizens. Aristotle well understood and anticipated this notion when he said that human beings could—and must—live in political communities if they are to live in accordance with their nature. And he defined politics in terms of creating a culture and social organization of individuals with a special kind of self-identity (citizenship), ruling themselves in common with equitable and just laws (*isonomia*), and seeking to achieve the human good together and the common good for all (*politia* or *res publica*). Citizenship for Aristotle was active, not passive. It consisted of ruling and being ruled in turn.

When viewed from this vantage point, it is clear that what we now call "political" activity in the world today is quite far removed from structures and value systems that can be expected to lead toward creative accommodation and governance that is just in its trusteeship over the good of the human and the biotic community. Our current politics cannot be the crucible for the reconciliation of humans and natural systems nor of accommodation to the functional limits of those systems. It cannot do so because it offers no counterpoint to the broader ethos and worldview of technological mastery of nature. That counterpoint is what I mean here by an understanding of the political. What we call politics today is not a bridle on the orientation of mastery but a handmaiden or extension of that orientation. The political is the counter-vision, the enabling act of mind, that will kick the habit of seeking material affluence rather than plenitude, and the habit of favoring competition, mastery, and extractive power over membership, care, and ruling and being ruled in turn.

Conclusion

In this book I have set out a notion of an ecological social contract, reflecting a relational moral ideal, a vision of citizenship as trusteeship, and an ecocentric cultural faith and ethos. I have also set out a vision of an ecological political economy, reflecting a governance that is discursively democratic and participatory, and economic thinking that reconceptualizes wealth as plenitude and frugality, property as a system of common-pool resources managed as a sustainable trust, and individual freedom as a lived relational practice with developmental goals of human and natural flourishing.

At the ethical core of this reconstruction of the social contract, which is in effect a new peace treaty with the planet, are the principles of right relationship (duties of care, concern, and respect for the integrity and flourishing of others) and right recognition (duties of equal moral consideration and concern for the rights of both human and nonhuman others as members in the shared moral community of life). The practical core of this reconstruction is the notion that widespread human action leading to personal, institutional, and cultural change rests on the twin motivations of conviction (reasoned conclusions concerning what is right and good) and contentment (a sensibility of fulfillment, meaning, and well-being). An ecological social contract is simply an interpretation of these principles and a fleshing out of the substantive content of these motivations that are ecological and relational.

I have drawn on, but not attempted to fully reproduce or summarize, an enormous body of scientific and philosophical research that warns us that our predominant systems of survival and modes of making a living on earth are neither naturally nor ethically sustainable. As a species today we extract and excrete more than our planet can tolerate well, or at least certain pockets and concentrations of the most powerful and affluent members of our species do so. This state of affairs I try to capture in the idea of the social contract of consumption and the consumptive political economy, a figure of speech in which a part (consumption) stands for and is meant to convey the essence of the whole. I seek an alternative to the consumptive society out of a deep concern that the survival of our social order, the health of the republic, and the hard-won freedoms and ideals of liberal and democratic enlightenment are in peril. This message is not new, to be sure. If anything, it has become so strident and frequently reiterated that it has a numbing, even nihilistic, effect on many people. My hope is to re-moralize, not to demoralize further.

Discussions of alternatives to the current status quo are also available in abundance, and they have a valuable tendency to be focused and practical. Renewable energy use, protection of wilderness, endangered species, and ecologically vulnerable areas, more sustainable agriculture, manufacturing, transportation, and building practices, and greenhouse gas emission reduction are among the most well developed of these alternative ideas worldwide. As an American, not being the world leader or first adopter of such innovations doesn't bother me, but not being more open to listen and learn from the wisdom and experience of others does. It makes me ashamed, and the fact that my country is being so insanely reluctant and slow to follow in the path of these innovations alarms me, for the sake of my newborn grandson, and for many other reasons besides. While so many thoughtful and creative people are working on solving problems, I turn my own attention to the complementary, and I think no less important, task of *solving for pattern*, as Wendell Berry put it. In this short book I have offered only glimpses and facets of a solved pattern, mere notes for a new peace treaty between humans and nature that are not fully orchestrated, but which do deserve a hearing as we assemble ourselves to create a new great charter, an ecological Magna Carta for the future.

Rhetorically my strategy in this book has drawn on an attempt to make the familiar strange. Only by directly confronting the central assumption of contemporary globalization, endless growth and material consumption, can we open a new space for ethical critique and the ecological imagination. Thinking otherwise is a precondition for living and doing otherwise. Many talk about why we deny climate change, and some even suggest that our brains are not hardwired by evolution in such a way as to allow us to think in the emotional and rational ways that are required today. On the other hand, there are conditions out of which energies for change in some of the directions reviewed in this book—plenitude rather than material affluence, richer civic lives and practical engagement, developing a deeper sense of meaningful craft work, public work, and the place of the commons in our world and in our lives—can be tapped. There is a widespread sense of exhaustion and frustration with *excess* of all kinds—consumptive, technological, financial.[1] The human sense of superiority over other living things is manifestly not being earned or lived up to as the sixth great extinction of life on earth proceeds apace.[2] The ubiquitous aspiration of mastery, which should be treated as a narcissistic fantasy, is instead treated as an accomplished fact that has reached the point of cultural satiation and disgust. Unsustainable growth in human extraction and excretion and human mastery over natural limits are not a sign of our freedom or our spiritual election, which the sociologist Max Weber discerned in the rise of capitalism, but a sign of our own domination and entrapment.[3]

Prefaced here, my ongoing concern in future work will be to find a form of democratic governance in a much more ecologically oriented world that remains liberal in its respect for human rights and the rule of law, but is more discursive and civic in its orientation than interest group liberalism has been. I believe this will require that we meet five key objectives in our conceptual work as intellectuals and in our discourse as citizens engaged in discursive democratic argument. We need to learn to talk about: (1) an ecological self rather than an extractive self—toward individuality rather than individualism—a new form of *identity*; (2) relational liberty rather than possessive negative liberty—a new form of *freedom*; (3) deliberation as distinct from bargaining—a new form of *governance*; (4) dialogic judgment as distinct from monologic assessment of interests—a new form of *reason*; and (5) practicing trusteeship instead of practicing consumerism—a new form of *living*.

We have much public work to do. Let's get on with it.

Acknowledgments

For thirty-five years I have been studying, writing, and teaching in the fields of moral philosophy, political theory, and bioethics. During the last decade, my work at the Center for Humans and Nature has focused my attention on philosophical and ethical issues concerning the human place in nature. That has led me toward a study of the interpretive understanding we have concerning our identity, our responsibilities, and our possibilities in our natural place, which, given our evolution, the capabilities we have, and the kind of creatures we are, is also our social place.

Those whose thinking and conversation have been invaluable to me in writing this book include the late Strachan Donnelley, who patiently over many years persuaded me of the importance of what I had been neglecting, as well as Richard Bernstein, Curt Meine, Ronald Engel, Joan Engel, Paul Heltne, Bruce C. Coull, George Rabb, Wes Jackson, George Raney, Gerald Adelmann, Elizabeth Minnich, Gregory Kaebnick, Brooke Hecht, Gavin Van Horn, Anja Claus, and Kate Cummings.

In struggling to get a purchase on the complex areas of ecological economics and ecological political economy, the expertise and guidance of the following colleagues have been enormously helpful and important to me: Benjamin Barber, Peter G. Brown, Richard Howarth, Gordon Burghardt, Lew Daly, James W. Fyles, Geoffrey Garver, Mark Goldberg, Carol

Gould, Janice Harvey, Richard Hayes, Richard Janda, Kathryn Kintzele, Stephen Latham, Qi Feng Lin, Richard Lehun, Jack P. Manno, Robert Nadeau, Nancy E. Mayo, Alexandre Poisson, Nicholas Robinson, Mark Schlesinger, Juliet B. Schor, David Smith, Gus Speth, Peter Timmerman, and Peter Victor.

In revising the text I have benefited greatly from the comments and suggestions of two anonymous reviewers and the supportive guidance of my editor, Derek Krissoff, of the West Virginia University Press. I am most grateful to my agent, Timothy Hays, for putting me in contact with him. Robert Nadeau and Michael Pellegrino read the manuscript carefully and gave me many helpful ideas and corrections, as did Ron Engel and Laura Berry.

The steadfast companionship of my wife, Maggie Jennings, over a lifetime has been the taproot of the reflections in this book. And as this book was being written, she and I were blessed by becoming grandparents for the first time. Our grandson's birth has given me a reconnection with the future. It is principally for him and for those in his generation that I write.

Notes

Epigraph

1. M. Serres, *The Natural Contract*, trans. E. MacArthur and W. Paulson (Ann Arbor: University of Michigan Press, 1995), 32, 38.

Introduction

1. P. G. Brown and G. Garver, *Right Relationship: Building a Whole Earth Economy* (San Francisco: Berrett-Koehler Publishers, 2009), 1.

2. J. Schor, *Plenitude: The New Economics of True Wealth* (New York: Penguin, 2010), 2.

3. N. Klein, *This Changes Everything: Capitalism vs. the Climate* (New York: Simon and Schuster, 2014), 21.

4. J. Hansen, P. Kharecha, M. Sato, V. Masson-Delmotte, F. Ackerman et al., "Assessing 'Dangerous Climate Change': Required Reduction of Carbon Emissions to Protect Young People, Future Generations and Nature," *PLoS ONE* 8, no. 12 (December 3, 2013): e81648, 20.

5. For a wry account of how the future may look back at our behavior in regard to climate change, see N. Oreskes and E. M. Conway, *The Collapse of Western Civilization: A View from the Future* (New York: Columbia University Press, 2014).

6. When I refer to "governance" in this book, I mean the overall process of coordinating, shaping, and directing individual and collective action. Governance involves government, but

also many other social institutions and cultural interpretations as well. For a fuller discussion of this notion, see Chapter 12, pp. 173–74.

7. B. McKibben, "The Pope and the Planet," *New York Review of Books*, August 12, 2015, 40.

8. Pope Francis, *Laudato Si': On Care for Our Common Home* (Huntington, IN: Our Common Visitor, 2015), §106.

9. One outstanding example, out of hundreds that could be cited, is A. Lovins and the Rocky Mountain Institute, *Reinventing Fire: Bold Business Solutions for the New Energy Era* (White River Junction, VT: Chelsea Green Publishing, 2011).

Part I: Rethinking Life, Liberty, and the Pursuit of Happiness on a Planet in Crisis

1. W. Churchill, "The Locus Years," in *Never Give In!: The Best of Winston Churchill's Speeches*, selected by W. S. Churchill (New York: Hyperion, 2003), 142–53.

2. D. Bonhoeffer, *Ethics*, trans. N. Smith (New York: Collier, 1965), 265.

1. The Social Contract

1. Republic 358e–359b, in *The Republic of Plato*, trans. A. Bloom (New York: Basic Books, 1968).

2. Quoted in N. O. Keohane, *Philosophy and the State in France: The Renaissance to the Enlightenment* (Princeton, NJ: Princeton University Press, 1980), 164.

3. J. Rawls, *A Theory of Justice*, rev. ed. (1971; Cambridge, MA: Harvard University Press, 1999). R. Nozick, *Anarchy, State, and Utopia* (New York: Basic Books, 1974). D. Gauthier, *Morals by Agreement* (New York: Oxford University Press, 1987). T. M. Scanlon, *What Do We Owe Each Other?* (Cambridge, MA: Harvard University Press, 2000). For an interesting discussion relating the social contract idea to contemporary ecology, see K. O'Brien, B. Hayward, and F. Berkes, "Rethinking Social

Contracts: Building Resilience in a Changing Climate,"
Economy and Society 14, no. 2 (2009): 1–17.

4. J.-J. Rousseau, *On the Social Contract*, Bk I, Ch. VI, in
R. D. Masters, ed., *On the Social Contract with Geneva
Manuscript and Political Economy* (New York: St. Martin's
Press, 1978), 53.

5. On the writing of the Declaration, see G. Wills, *Inventing
America: Jefferson's Declaration of Independence* (Garden
City, NY: Doubleday, 1978).

6. For accessible and thoughtful accounts, see B. R. Barber,
*Consumed: How Capitalism Corrupts Children, Infantilize
Adults, and Swallow Citizens Whole* (New York: W. W.
Norton, 2007); R. B. Reich, *Supercapitalism: The Transforma-
tion of American Business, Democracy, and Everyday Life*
(New York: Random House, 2007).

7. Oxfam International, *Working for the Few: Political Capture
and Economic Inequality*, Briefing Paper, January 20, 2014,
https://www.oxfam.org/en/research/working-few.

2. Political Economy

1. See, for example, J. Asafu-Adjaye, L. Blomqvist, S. Brand,
B. Brook, R. Defries, E. Ellis et al., "An Ecomodernist
Manifesto," April 2015, www.ecomodernism.org. This
interesting and widely discussed statement, written by those
associated with the Breakthrough Institute, cofounded by Ted
Nordhaus and Michael Shellenberger, differs from my argument in this book in several respects. Most important, it does
not address the role of political economy and has little to say
about governance as an institutional structure, as distinct from
specific policy proposals and ideas. I regard both as mediat-
ing structures that shape the kinds of ecological burdensome
pressure that human activity puts on other systems of life.

However, the manifesto does engage with a fundamental
question concerning how best to articulate and understand
the possible and appropriate relationship between nature and
human beings. The eco-modernist perspective is characterized

by the recognition of the need to lessen and make more sustainable human impacts on the planet, and by the rejection of the notion that in order to do so, humanity must "harmonize" itself with nature and natural realities that are given and constraining. The emphasis instead is on making humanity less dependent on nature, more detached from natural limits, principally via technological innovation. In this way the goals of human progress, well-being, and liberation can be achieved by decoupling these goals from activities that have a deleterious impact on natural systems to the extent possible.

The technological cocoon envisioned here would be one solution to the conundrum posed by the social contract of consumption. But they do not so much propose a solution to the problem as facilely erase one of the premises that creates the dilemma in the first place. An ecologically low-impact, or naturally free-standing technology, resting on virtually unlimited energy from the sun or from the controlled manipulation of subatomic forces (such as nuclear fusion), would indeed provide a way out, but at this point it remains on a par with the advice of the economist in a deep hole: assume a ladder. If harmonizing the human with nature is too humanly self-effacing, then decoupling ourselves from our dependence—or better interdependence—with nature is too humanly self-absorbed.

The solution to both forms of misprision, in my view, is not a simple formulation by any means, but it does require a sense of responsibility and trusteeship that is ontologically entangled rather than decoupled. The complex formulation I seek also requires a use of technology that is not aiming for the artificial as a substitute for the natural, but for a human creativity, mediated by tools, that is applied *with* the grain of natural material. Not against its grain, as is so often done now, but then not *outside* its grain, as the manifesto seemingly would have it, either. For a thoroughgoing and thought-provoking treatment of these and related issues, see: J. Purdy,

"The Politics of Nature: Climate Change, Environmental Law, and Democracy," *The Yale Law Journal* 119 (2010): 11–22–1209, and J. Purdy, "Our Place in the World: A New Relationship for Environmental Ethics and Law," *Duke Law Journal* 62, no. 4 (2013): 857–932.

2. K. Boulding, "Earth as a Space Ship" (1965): 1–2, http://earthmind.net/earthmind/docs/boulding-1965.pdf.

3. See T. Piketty, *Capital in the Twenty-First Century* (Cambridge, MA: Harvard University Press, 2014).

4. P. A. Victor, *Managing Without Growth: Slower by Design, Not Disaster* (Cheltenham, UK: Edward Elgar, 2008), 27–29, 32.

5. J. Rockström, J. W. Steffen, K. Noone, Å. Persson, F. S. Chapin III, E. Lambin, T. M. Lenton, M. Scheffer, C. Folke, H. Schellnhuber, B. Nykvist, C. A. De Wit, T. Hughes, S. van der Leeuw, H. Rodhe, S. Sörlin, P. K. Snyder, R. Costanza, U. Svedin, M. Falkenmark, L. Karlberg, R. W. Corell, V. J. Fabry, J. Hansen, B. Walker, D. Liverman, K. Richardson, P. Crutzen, and J. Foley, "Planetary Boundaries: Exploring the Safe Operating Space for Humanity," *Ecology and Society* 14, no. 2 (2009): 32, http://www.ecologyandsociety.org/vol14/iss2/art32/.

   An updated version of this important work is W. Steffen, K. Richardson, J. Rockström, S. E. Cornell, O. Fetzer, E. M. Bennett, R. Biggs, S. R. Carpenter, W. de Vries, C. A. de Wit, C. Folke, D. Gerten, J. Heinke, G. M. Mace, L. M. Persson, V. Ramanathan, B. Reyers, and S. Sörlin, "Planetary Boundaries: Guiding Human Development on a Changing Planet," *Science* 347, no. 1259855 (2015), DOI: 10.1126/science.1259855.

6. P. Davies, *The Goldilocks Enigma: Why Is the Universe Just Right for Life?* (New York: Mariner Books, 2008).

7. I. Berlin, "Two Concepts of Freedom," in *Four Essays on Liberty* (Oxford, UK: Oxford University Press, 1969), 131.

8. For Arendt, action was the facet of the human condition associated with political life, the *polis* or public realm, while

labor and work were at home in the household, the *oikos* or the private economic and reproductive realm. In my view, her dualism in this regard is problematic and need not be accepted in order to gain insight from her account of the three dimensions of humanness. Indeed, ecological economics suggests that labor, work, and action are dimensions of all realms of social life and that the differentiation between public and private—politics and economics, *polis* and *oikos*—is not the ethically desirable ideal that Arendt sometimes made it out to be. See S. Benhabib, *Situating the Self* (New York: Routledge, 1992), 89–120.

Part II: Natural Being, Cultural Becoming: Nature in Humans

1. J.-J. Rousseau, *Discourse on the Origins and Foundations of Inequality among Men*, in R. D. Masers and C. Kelly, eds., *The Collected Writings of Rousseau*, vol. 3 (Hanover, NH: The University Press of New England, 1992), 37–38.

2. J. Dewey, "Creative Democracy—The Task Before Us," in *The Political Writings*, ed. D. Morris and I. Shapiro (1939; Indianapolis: Hackett, 1993), 241.

3. The Roots and Logic of Social Contract Theory

1. This chapter and the next extend the reading of Hobbes and Rousseau that I present in B. Jennings, "Nature as Absence: The Natural, the Cultural, and the Human in Social Contract Theory," in G. Kaebnick, ed., *The Ideal of Nature: The Appeal to Nature in Debates about Biotechnology and the Environment* (Baltimore: Johns Hopkins University Press, 2011), 29–48.

2. The third most important and influential social contract writer was the English philosopher John Locke (1632–1704). Locke came about twenty to thirty years after Hobbes and sixty years before Rousseau, and he was active in the 1680s and 1690s. His version of contractarianism set out to refute Robert Filmer and the divine right of kings (the first part of his main work of political theory, *Two Treatises of*

Government) and to establish through moral consent a kind of sovereign authority that was not absolute (the second part of *Two Treatises of Government*). Locke wrote to defend the Glorious Revolution of 1688–89, not after the fact, as was long thought, but before the revolution that brought King William of Orange and Queen Mary to England and replaced the Stuart dynasty with the House of Hanover. (Locke's *Two Treatises of Government* was published anonymously, and its dates of composition were only correctly established by scholarship in the 1950s. (P. Laslett, "Introduction," in John Locke, *Two Treatises of Civil Government*, 3rd ed. (1963; Cambridge, UK: Cambridge University Press, 1988), 3–127.) It also paved the way for the establishment of a much more limited, constitutional monarchy that shared power with Parliament. These are hallmarks in the development of liberal democracy, and Locke was the great theorist of them.

Locke's sovereign, which he did not identify with the monarch or the aristocracy, but with the majority of the people (i.e., those males who owned property and had a right to vote), does not need to be as unlimited or powerful as Hobbes's because for him the natural condition is not as dire. Locke's state of nature has two stages. The first stage is a scene of personal freedom and self-sovereignty but checked by natural reason that led most men to be peaceable and cooperative most of the time. Locke's state of nature does not lack authority or morality, but does lack impartial judges and interpreters of common rules for living. The rules themselves are moral, unwritten, and rather vague. But because there is abundance in physical nature, a reasonable basis for limited trust and secure future planning, and natural fairness, the first stage of the state of nature is able to do without absolute sovereignty or, indeed, without government of any kind. In the second stage of the state of nature, conflict breaks out due to developing inequality and artificial scarcity. Locke attributes this to the invention of money.

Before money, in a natural barter economy, individuals appropriate only as much as they could use without spoilage. There is no way to stockpile and preserve natural consumable goods, and because anyone could easily return to the natural stockpile at any time, there is no self-interested reason to hoard or to try to seize another's resources. When goods could be transformed into a durable form of money, unlimited accumulation became reasonable and, in fact, highly desirable. This leads to conflict and to the establishment of government in order to provide impartial arbitrators to resolve disputes, to set rules for the protection of private property and the promotion of commerce, and to manage production and distribution in the political economy. Locke paid much more attention to the exploitation and appropriation of physical nature or environmental resources than did Hobbes and Rousseau.

3. The scholarly literature on Hobbes is voluminous. Among the works that have shaped my own interpretation and are pertinent to the ways in which his thought figures into the concerns of this book, see N. Malcolm, *Aspects of Hobbes* (Oxford, UK: Oxford University Press, 2002); Q. Skinner, *Reason and Rhetoric in the Philosophy of Hobbes* (Cambridge, UK: Cambridge University Press, 1996); R. Polin, *Politique et Philosophie chez Thomas Hobbes*, 2nd ed. (Paris: J. Vrin, 1977); and C. Schmitt, *The Leviathan in the State Theory of Thomas Hobbes: Meaning and Failure of a Political Symbol*, trans. G. Schwab and E. Hilfstein (Chicago: University of Chicago Press, 2008).

4. Classic statements of this mixed constitutional arrangement can be found in A. Hamilton, J. Madison, and J. Jay, *The Federalist Papers* (New York: Signet, 2003), nos. 10 and 51, and in C. de Montesquieu, *The Spirit of the Laws*, ed. A. M. Cohler, B. C. Miller, and H. S. Stone (Cambridge, UK: Cambridge University Press, 1989).

5. Like that on Hobbes, the scholarship on Rousseau is rich and varied. Such thinkers do not lend themselves to final or

definitive interpretations; each generation of readers tends to find what it needs for its time and place in their texts. My reading of Rousseau has been influenced strongly by R. Polin, *La Politique de la Solitude: Essai sur J.-J. Rousseau* (Paris: Editions Sirey, 1971), and J. Starobinski, *Jean-Jacques Rousseau: Transparency and Obstruction*, trans. A. Goldhammer (Chicago: University of Chicago Press, 1988).

6. It should be noted that Rousseau was a significant naturalist, in addition to being a social philosopher, novelist, playwright, composer, and autobiographer. He wrote widely on botany and on various landscapes, stressing the moral and spiritual effects of living within environments of undisturbed natural beauty.

7. This observation may be controversial when it comes to Locke, but not, I think, for Hobbes and Rousseau. For a discussion of the importance of religion in Locke, see J. Dunn, *The Political Thought of John Locke* (Cambridge, UK: Cambridge University Press, 1969).

8. Rousseau, *On the Social Contract*, Bk. I, Ch. VIII, 55–56.

9. The defining features of "the natural" in social contract theory are not those that have been emphasized in connection with "the primitive" or "the savage" in nineteenth-century social anthropology—preliterate communication, nonabstract and nonscientific or animistic modes of thought, technologically simple hunting and gathering economies with small band organization and little impact on the ecosystems they inhabit beyond the carrying capacity of those systems. See R. L. Meek, *Social Science and the Ignoble Savage* (New York: Cambridge University Press, 1976); M. Diamond, *In Search of the Primitive* (New Brunswick, NJ: Transaction, 1974); R. Clarke and G. Hindley, *The Challenge of the Primitives* (New York: McGraw-Hill, 1975); and R. Wokler, *Rousseau: A Very Short Introduction* (New York: Oxford University Press, 2001).

10. M. Sahlins, *Culture and Practical Reason* (Chicago: University of Chicago Press, 1976), and C. Geertz, *The Interpretation of Cultures* (New York: Basic Books, 1973).

11. Geertz, *The Interpretation of Cultures*, ix.

12. D. Korten, *The Great Turning: From Empire to Earth Community* (San Francisco: Berrett-Koehler, 2006), 756.

13. The distinction between a drive and an instinct is important. See J. Laplanche and J.-B. Pontalis, *The Language of Psychoanalysis* (London: The Hogarth Press, 1973), 214–16.

14. Rousseau, *Discourse on the Origins and Foundations of Inequality*, 47–58.

15. The notion of the Golden Age was well established in Ovid and other classical sources and was a familiar rhetorical convention in Rousseau's time. Among many examples that could be cited, see the remarkable depiction of the Golden Age in M. de Cervantes Saavedra, *Don Quixote*, trans. E. Grossman (New York: HarperCollins, 2003), Pt 1. Ch. 11, 76–78.

4. The Uses of Nature and Culture: Artifice and Accommodation

1. T. Hobbes, *Leviathan*, ed. M. Oakeshott, Pt. 1, Ch. 11 (Oxford, UK: Basil Blackwell, 1946), 64.

2. J.-J. Rousseau, *Confessions*, trans. J. M. Cohen (Harmondsworth: Penguin, 1954), 19.

3. Hobbes, *Leviathan*, Pt. 1, Ch. 8, 46.

5. Re-enchanting the Social Contract

1. Hume, by the way, took a dim view of social contract theory, regarding it as a false historical account of the origin of government. See D. Hume, "Of the Original Contract," in *Essays: Moral, Political, and Literary* (New York: Cosimo Classics, 2006), 452–73. Hume does not belong to the contractarian tradition but instead paved the way for the development of the ethical theory of utilitarianism. He

loathed Hobbes's political absolutism, but was personally kind to Rousseau, whom he tried to protect from persecution in France by arranging a safe haven in England. But Rousseau, exceedingly emotionally volatile, was a bad fit with the affable Hume and came to distrust his motives. Ultimately Rousseau did not accept the largesse that Hume had arranged.

2. P. Anderson, *Passages from Antiquity to Feudalism* (London: New Left Books, 1974). P. Anderson, *Lineages of the Absolutist State* (London: New Left Books, 1974). E. M. Woods, *Citizens to Lords: A Social History of Western Political Thought from Antiquity to the Middle Ages* (London: Verso, 2008). E. M. Woods, *Liberty and Property: A Social History of Western Political Thought from the Renaissance to the Enlightenment* (London: Verso, 2012).

3. H. S. Maine, *Ancient Law* (New York: E. P. Dutton, 1917). F. Tönnies, *Community and Society* (New York: Harper and Row, 1963). B. Nelson, *The Idea of Usury: From Tribal Brotherhood to Universal Otherhood*, 2nd ed. (Princeton, NJ: Princeton University Press, 1969). C. B. Macpherson, *The Political Theory of Possessive Individualism: Hobbes to Locke* (New York: Oxford University Press, 1962).

4. J. Osterhammel, *The Transformation of the World: A Global History of the Nineteenth Century*, trans. P. Camiller (Princeton, NJ: Princeton University Press, 2014), 637–73.

5. On this general perspective see J. G. A. Pocock, *Politics, Language, and Time: Essays on Political Thought and History* (New York: Athenaeum, 1973). For specific contextual studies, see Q. Skinner, *Hobbes and Republican Liberty* (Cambridge, UK: Cambridge University Press, 2008). R. Ashcraft, *Revolutionary Politics and Locke's Two Treatises of Government* (Princeton, NJ: Princeton University Press, 1986). N. O. Keohane, *Philosophy and the State in France: The Renaissance to the Enlightenment* (Princeton, NJ: Princeton University Press, 1980). C. Blum, *Rousseau and*

the Republic of Virtue: The Language of Politics in the French Revolution (Ithaca, NY: Cornell University Press, 1986).

6. R. L. Nadeau, *Rebirth of the Sacred: Science, Religion, and the New Environmental Ethos* (New York: Oxford University Press, 2012).

7. This is analogous to the population-balancing mechanism between predator species and prey species that has been described in the evolutionary and ecological sciences. Indeed, in the nineteenth century, and persisting today, there are many parallels between conceptual frameworks in the fields of economics and evolutionary biology.

8. E. D. Larson, "Can U.S. Carbon Emissions Keep Falling?" *Climate Central*, October 2012, http://www.climatecentral.org /wgts/can-emissions-keep-falling/CanEmissionsKeepFalling .pdf.

9. I borrow the term "strong ontology" and key ideas in this section from S. K. White, *Sustaining Affirmation: The Strengths of Weak Ontology in Political Theory* (Princeton, NJ: Princeton University Press, 2000). See also P. Hoffman, *Freedom, Equality, Power: The Ontological Consequences of the Political Philosophies of Hobbes, Locke, and Rousseau* (New York: Peter Lang, 1999).

10. See J. Rawls, *Political Liberalism* (New York: Columbia University Press, 1993). For a general discussion of this issue, see S. Mulhall and A. Swift, *Liberals and Communitarians*, 2nd ed. (Oxford, UK: Blackwell, 1996).

11. J. Jacobs, *Systems of Survival: A Dialogue on the Moral Foundations of Commerce and Politics* (New York: Random House, 1992).

12. A. Hirschman, *The Passions and the Interests: Arguments for Capitalism Before Its Triumph* (Princeton, NJ: Princeton University Press, 1977).

13. J. Israel, "How the Light Came In" (review of A. Pagden, *The Enlightenment*), *The Times Literary Supplement*, June 21, 2013, 9–10.

14. See R. A. Dahl, *Polyarchy: Participation and Opposition* (New Haven, CT: Yale University Press, 1971), and L. H. Gunderson and C. S. Holling, eds., *Panarchy: Understanding Transformations in Human and Natural Systems* (Washington, DC: Island Press, 2001).

15. A. Tocqueville, *Democracy in America*, ed. and trans. H. C. Mansfield and D. Winthrop (Chicago: University of Chicago Press, 2000), 662–63.

16. C. Taylor, *The Ethics of Authenticity* (Cambridge, MA: Harvard University Press, 1991), 112–13, 118.

17. C. Taylor, *A Secular Age* (Cambridge, MA: Harvard University Press, 2007).

18. C. Taylor, "A Catholic Modernity?" in *Dilemmas and Connections: Selected Essays* (Cambridge, MA: Harvard University Press, 2011), 167–87.

19. Rousseau, *Discourse on the Origins and Foundations of Inequality*, 36.

20. Ibid., 91. As the passage continues, Rousseau attributes the birth of *amour propre* to the development of self-consciousness.

21. J.-J. Rousseau, *Émile or On Education*, trans. A. Bloom (New York: Basic Books, 1979), 213–14.

22. Rousseau, *Discourse on the Origins and Foundations of Inequality*, 66.

23. Ibid., 36.

Part III: Terms of an Ecological Contract: Humans in Nature

1. J. S. Mill, *Principles of Political Economy in Collected Works*, ed. J. M. Robson, vols. 2 and 3 (Toronto: University of Toronto Press, 1965), 3:756–57.

2. H. Arendt, *On Violence* (New York: Harcourt, Brace, and World, 1970), 82–83, 84.

3. W. C. Williams, "The Orchestra," in *Pictures from Breughel and Other Poems* (New York: New Directions, 1962), 82.

6. Agency, Rules, and Relationships in an Ecological Social Contract

1. M. C. Nussbaum, *Creating Capabilities: The Human Development Approach* (Cambridge, MA: Harvard University Press, 2011).
2. S. Weil, "Reflections on the Right Use of School Studies with a View to the Love of God," in *Simone Weil Reader*, ed. G. A. Panichas (Mt. Kisco, NY: Moyer Bell Ltd., 1977), 51.

7. Wealth: From Affluence to Plenitude

1. Schor, *Plenitude*, 2.
2. E. D. Schneider and D. Sagan, *Into the Cool: Energy Flow, Thermodynamics, and Life* (Chicago: University of Chicago Press, 2006).
3. Y.-M. Abraham, "Little Vade Mecum for the Growth Objector," May 2011, http://montreal.degrowth.org/aboutdegrowth.html.
4. S. Latouche, *Farewell to Growth* (Cambridge, UK: Polity Press, 2009), and S. Latouche, "De-growth," *Journal of Cleaner Production* 18 (2010): 519–22; G. Kallis, "In Defense of Degrowth," *Ecological Economics* 70, no. 5 (March 15, 2011): 873–80; J. U. Martinez-Alier, F.-D. Vivien Pascual, and E. Zaccai, "Sustainable De-growth: Mapping the Context, Criticisms, and Future Prospects of an Emergent Paradigm," *Ecological Economics* 69 (2010): 1741–47.
5. J. E. Stiglitz, A. Sen, and J.-P. Fitoussi, *Mismeasuring Our Lives: Why GDP Doesn't Add Up* (New York: The New Press, 2010).

8. Property: From Commodity to Commons

1. Rousseau, *Discourse on the Origins and Foundations of Inequality*, 43.
2. A. Leopold, *A Sand County Almanac and Sketches Here and There* (1949; New York: Oxford University Press, 1989), vii.

3. J. Waldron, *The Right to Private Property* (Oxford, UK: Oxford University Press, 1988), 31.

4. L. von Mises, *Human Action: A Treatise on Economics* (New York: Laissez Faire Books, 2008), 652.

5. K. Polanyi, *The Great Transformation: The Political and Economic Origins of Our Time* (Boston: Beacon Press, 1957), 73.

6. See B. Nelson, *The Idea of Usury: From Tribal Brotherhood to Universal Otherhood*, 2nd ed. (Princeton, NJ: Princeton University Press, 1969).

7. For an interesting study of this tension, see D. Bell, *The Cultural Contradictions of Capitalism* (New York: Basic Books, 1976).

8. K. Marx, *Capital*, vol. III (New York: Penguin Books, 1981), 911.

9. Polanyi, *The Great Transformation*, 178.

10. A. O. Lovejoy, *The Great Chain of Being: A Study of the History of an Idea* (Cambridge, MA: Harvard University Press, 1936). C. Taylor, *Hegel* (Cambridge, MA: Cambridge University Press, 1975), 3–50. E. J. Dijksterhuis, *The Mechanization of the World Picture* (Oxford, UK: Oxford University Press, 1961). E. A. Burtt, *The Metaphysical Foundations of Modern Physical Science*, 2nd rev. ed. (Garden City, NY: Doubleday, 1954). S. Gaukroger, *The Emergence of a Scientific Culture: Science and the Shaping of Modernity 1210–1685* (Oxford, UK: Oxford University Press, 2006).

11. See M. J. Sandel, *What Money Can't Buy: The Moral Limits of Markets* (New York: Farrar, Straus and Giroux, 2012).

12. On this notion, see J. Searle, *Speech Acts: An Essay in the Philosophy of Language* (Cambridge, UK: Cambridge University Press, 1969).

13. The term "land" is a synecdoche (a figure of speech in which a part represents the whole) in Leopold's writings: it stood for the biotic community in a broad sense—soil, the watershed, ecosystemic interactions between organic and inorganic systems and energy exchanges, and the like.

14. Note that it is grammatically necessary, albeit awkward, in English to refer to "the public" as a noun and not merely an adjective modifying something else; or else one must resort to metaphors such as "the public space" or "the public sphere."

This grammatical nominalization is one source of the tendency to reify this concept, a danger to which I will return.

15. M. G. Dietz, "'The Slow Boring of Hard Boards': Methodical Thinking and the Work of Politics," *American Political Science Review* 88, no. 4 (December 1994): 873–86.

16. S. S. Wolin, "Political Theory: Trends and Goals," in D. S. Sills, ed., *International Encyclopedia of the Social Sciences* (New York: Macmillan Co., 1968), 12:318–31.

17. The distinction between the "transactional" and the "transformative" was suggested to me by J. M. Burns, *Leadership* (New York: Harper and Row, 1978). It parallels the distinction I make between the language of political bargaining and political discourse or deliberation. Transactional relationships coordinate and manage the interests, beliefs, and emotions that the individual parties to the relationship bring to it from the beginning. They take what is given and organize it so as to produce valued and beneficial results. Transformational relationships may begin with transactional intent, but through the process of cooperation and communication are gradually transformed into experiences and practices that are more developmental and open-ended. The parties to these relationships may change in terms of their understanding, motivation, and capabilities in and through these relationships.

With this distinction in mind, in earlier work on the thesis and argument of this book, I chose to contrast a social contract with a social covenant because the idea of a contract denotes only a transactional approach and horizon. (See B. Jennings, "Beyond the Social Contract of Consumption," *Critical Policy Studies* 4 no. 3 (October 2010): 222–33.) Indeed, there is much to be said for pressing this distinction forward and there is a great need for serious work on the concept of covenant in environmental ethics and elsewhere. (See R. J. Engel, "Property: Faustian Pact or New Covenant with Earth?" in D. Grinlinton and P. Taylor, eds., *Property Rights and Sustainability: The Evolution of Property Rights to Meet Ecological Challenges*

(Leiden and Boston: Martinus Nijhoff Publishers, 2011), 63–86.)
For this work, however, I did not feel that I could do justice to
the concept of covenant and its differences from the concept of
contract, so I decided to limit myself to the term "contract" and
to use it in an open-ended way. A substantive version of the
social contract, such as the consumptive contract, may be
limited to transactional functions only. But I explore the
possibility that a substantive version of the ecological social
contract can open onto a transformative process.

18. G. Hardin, "The Tragedy of the Commons," *Science* 162
     (December 13, 1968): 1243–48.

19. E. Ostrom, *Governing the Commons: The Evolution of
     Institutions for Collective Action* (Cambridge, UK: Cambridge
     University Press, 1990), 1.

20. The first and second solutions, the Hobbesian sovereign and
     the private property/market price system, do not alter the
     herdsmen's private identity as entrepreneurs, they merely
     change their incentive structures.

21. See E. Ostrom, *Governing the Commons, The Future of the
     Commons: Beyond Market Failure and Government Regula-
     tion* (London: Institute of Economic Analysis, 2012), and
     E. Ostrom, C. Chang, M. Pennington, and V. Tarko, *The
     Future of the Commons* (London: Institute of Economic
     Affairs, 2012). See also Y. Benkler, *The Wealth of Networks:
     How Social Production Transforms Markets and Freedom*
     (New Haven, CT: Yale University Press, 2006). D. Bollier,
     *Think Like a Commoner: A Short Introduction to the Life of the
     Commons* (Gabriola Island, BC: New Society Publishers,
     2014). D. Wall, *The Commons in History: Conflict, Culture,
     and Ecology* (Cambridge, MA: MIT Press, 2014).

9. Freedom: Relational Interdependence

1. Domination is a state of such narrow and thoroughgoing
    closure in one's life that it negates the purposive agency of the
    subject, and, at an extreme, the underlying capability to

exercise agency that could be deemed the person's own or to even conceive of oneself as a subject of agency. Domination must be distinguished from determination. Freedom from domination is compatible with the scientific theories and explanations that identify the determining conditions of human behavior, to which ecological economics appeals. Freedom from domination is freedom from arbitrary and contingent social and psychological determinants of behavior (and also of thought). This aspiration, which is in the modern world a moral imperative, does not require the abstract fantasy of an organically disembodied and socially disembedded individual.

2. I borrow the terms "extractive liberty" and "possessive individualism" from the work of C. B. Macpherson. See *Democratic Theory: Essays in Retrieval* (Oxford, UK: Oxford University Press, 1973), 118, and *The Political Theory of Possessive Individualism: Hobbes to Locke* (Oxford, UK: Oxford University Press, 1962).

3. For certain purposes it is useful to differentiate the concepts of freedom and liberty. Freedom often has a more private or personal connotation, while liberty has a resonance that links it more specifically to the political realm. Moreover, the idea of freedom is often associated with the philosophical issues of determinism and freedom of the will in a metaphysical sense. Here my focus is on moral and political liberty and the relationship between these normative ideals and the conditions of human agency and responsibility. If one is free or at liberty in this sense, one can reasonably be said to be the author of one's own actions and to bear responsibility for them. Liberty therefore has to do both with the will and motivation of an individual and with the social capacities and relationships available to an individual as they bear on the reasonable preconditions and ascription of moral responsibility for action. In this I follow P. Pettit, *A Theory of Freedom: From the Psychology to the Politics of Agency* (New York: Oxford University Press, 2001). For a discussion of the differences between the two concepts, see H. F. Pitkin, "Are

Freedom and Liberty Twins?" *Political Theory* 16, no. 4 (November 1988): 523–52.

4. The discussion in this chapter draws on B. Jennings, "Ecological Political Economy and Liberty," in P. G. Brown and P. Timmerman, eds., *Ecological Economics for the Antropocene: An Emerging Paradigm* (New York: Columbia University Press, 2015), 272–317.

5. It is worth emphasizing that the idea and ideal of negative liberty is inseparable from an atomistic conception of the unencumbered self and from a form of agency that involves the exercise of extractive power in the interactions between the self and others, and between human beings and the natural world. This conception of liberty is deeply ingrained in contemporary political culture, quintessentially in the United States, but increasingly throughout the global North as a whole.

6. J. S. Mill, *On Liberty* (1859; Indianapolis: Bobbs Merrill, 1956). J. A. Schumpeter, *Capitalism, Socialism, and Democracy*, 3rd ed. (New York: Harper and Row, 1950).

7. Macpherson, *Democratic Theory*, 18–19.

8. See W. Gaylin and B. Jennings, *The Perversion of Autonomy: The Uses of Coercion and Constraint in a Liberal Society*, 2nd ed. (Washington, DC: Georgetown University Press, 2003), and C. Taylor, *The Ethics of Authenticity* (Cambridge, MA: Harvard University Press, 1991).

9. J. Nedelsky, *Law's Relations: A Relational Theory of Self, Autonomy, and Law* (New York: Oxford University Press, 2013); K. J. Gergen, *Relational Being: Beyond Self and Community* (New York: Oxford University Press, 2009). See also C. Hamilton "Consumerism, Self-Creation, and Prospects for New Ecological Consciousness," *Journal of Cleaner Production* 18 (2010): 571–75.

10. In this discussion I am drawing on and developing a view that I have elsewhere explored in the context of related issues in the field of public health. Compare B. Jennings, "Public Health and Civic Republicanism," in A. Dawson and M. Verweij, eds., *Ethics, Prevention, and Public Health* (Oxford,

UK: Oxford University Press, 2007), 30–58; B. Jennings, "Public Health and Liberty," *Public Health Ethics* 2, no. 2 (July 2009): 123–34; and B. Jennings, "Relational Liberty Revisited: Membership, Solidarity, and a Public Health Ethics of Place," *Public Health Ethics* 8, no. 1 (February 2015): 7–17.

11. J. A. Schumpeter, *Capitalism, Socialism, and Democracy*, 3rd ed. (New York: Harper and Row, 1950).

12. R. Sennett, *The Craftsman* (New Haven, CT: Yale University Press, 2009).

13. C. Taylor, *Sources of the Self: The Making of Modern Identity* (Cambridge, MA: Harvard University Press, 1989), and R. Harré, *The Singular Self* (London: Sage Publications, 1998).

14. It is unfortunate that economists have either ignored this dialectical perspective on meaningful, intentional agency or have rejected it in favor of models of strategic action, rational gaming, and choice that are actually not supported by historical and social scientific evidence. See I. Shapiro and D. Green, *Pathologies of Rational Choice Theory* (New Haven: Yale University Press, 1994).

15. H. E. Daly and J. M. Cobb Jr., *For the Common Good: Redirecting the Economy Toward Community, the Environment, and a Sustainable Future*, 2nd ed. (Boston: Beacon Press, 1994).

16. In my view, citizenship does not have some external state of affairs called the "common good" as its instrumental objective. The common good is constituted by the proper institutionalization and functioning of citizenship and by the proper embedding of communal and ecological responsibility in the lifeworld. The common good is not a notion that sets up a test for particular policies or particular actions to meet (as does the parallel concept of "the public interest" in utilitarianism or liberal welfareism). It is not an outcome or an effect. However, the notion of the common good does provide a touchstone for judging and appraising a particular policy or decision. It appraises policy against criteria such as nondomination, nonarbitrariness, reasonable authority, mutual respect, reciprocity, and equity.

Contemporary utilitarianism tends to define interests or "utilities" abstractly across a population of individuals who have, as it were, only external or instrumental relationships to those interests. Utilitarianism also tends to ignore the distributional patterns in which these interests are fulfilled or their impact on discrete individual persons as such; it focuses instead on the net maximization of satisfaction or interest fulfillment in the aggregate. See L. Robbins, *An Essay on the Nature and Significance of Economic Science* (London: Macmillan, 1962), and V. Walsh, *Rationality, Allocation, and Reproduction* (Oxford, UK: Clarendon Press, 1996). By contrast, the judgments that make up communitarian policy appraisal are judgments of fittingness, character, and appropriateness. They must take into consideration the conditions of power and meaning that constitute the identity and interests of each person as a unique individual. They are at the political boundary between moral and aesthetic judgment. As such, they cannot be the *only* means of policy appraisal in economic policy, regulation, or law. But neither should they be left out altogether. See M. C. Nussbaum, *Poetic Justice* (Boston: Beacon Press, 1995), and K. Günther, *The Sense of Appropriateness: Application Discourses in Morality and Law* (Albany: State University of New York Press, 1993).

Shared purposes or problems are not the same as individual purposes or problems that happen to overlap for large numbers of people. Of course, they do affect persons as individuals and as members of smaller groups, but they also affect the constitution of a "people," a population of individuals as a structured social whole. An aggregation of individuals becomes a people, a public, a political community when it is capable of recognizing common purposes and problems in this way; when it achieves a certain kind of political imagination. See B. Anderson, *Imagined Communities: Reflections on the Origin and Spread of Nationalism*, 2nd ed. (New York: Verso, 1991).

17. A. Honneth, *The Struggle for Recognition: The Moral Grammar of Social Conflicts* (Cambridge, MA: MIT Press, 1996).

18.   G. Burghardt, *The Genesis of Animal Play* (Cambridge, MA: MIT Press, 2006).

10. Citizenship: From Electoral Consumer to Ecological Trustee

1.   The mainstream framework of democratic governance consists of interest group pluralism and representative democracy. Within this framework, democratic institutions are responsive to individual interests, concatenated or organized by the formation of various group structures that compete for the attention of popularly elected officials. Their competition in this regard consists both of the marketplace of ideas and the marketplace of campaign contributions, and other financial incentives for public officials. Modern societies are too large and complex to be governed by direct participatory mechanisms; democracy consists essentially in the right to vote and free and fair competition among candidates and parties for the support of self-interested voters. The political theory of interest group democracy was classically developed by James Madison, in *The Federalist*, especially nos. 10 and 51. The best contemporary explication and defense of this theory is R. A. Dahl, *A Preface to Democratic Theory* (Chicago: University of Chicago Press, 1956). Important critiques can be found in P. Bachrach, *The Theory of Democratic Elitism: A Critique* (Boston: Little Brown, 1967), and T. J. Lowi, *The End of Liberalism*, 2nd ed. (New York: Norton, 1979). In his notion of "juridical democracy" Lowi may be said to prefigure aspects of what I am calling ecological constitutionalism.

Interest group democracy as a type of political theory bears a striking resemblance to the orientation of mainstream economic and market theory, and no wonder, because for at least fifty years there has been much cross-fertilization between the two fields, so much so that many now consider political science to be a subfield of economics. One of the most influential works of political science in the last century was A. Downs, *An Economic Theory of Democracy* (New York: Harper and Row, 1957). It is interesting to compare Downs with Dahl's *Preface*, which was published at virtually the same time.

2. The principal alternative to interest group liberal democracy is a mode of democratic governance generally known as deliberative democracy or sometimes "discursive" democracy. It differs from interest group democracy in some fundamental respects. It challenges the primacy of a rather utilitarian and materialistic notion of interests as the basis of the psychological and moral dimension of a democratic polity. It also argues for both the feasibility and the normative justification of a more participatory form of democratic citizenship. Not passive electoral consumers, but active democratic trustees. However, it is not clear how deeply the theorists of deliberative democracy challenge some of the fundamental assumptions about individualism, autonomy, and the aggregative nature of social utility, the public interest, or the common good. Whether deliberative democracy is the appropriate governance model to correspond to ecological economics and to the new covenant of ecological trusteeship remains to be seen.

3. C. Taylor, "What's Wrong with Negative Liberty," in *Philosophy and the Human Sciences, Philosophical Papers*, vol. 2 (Cambridge, UK: Cambridge University Press, 1985), 210–29.

4. Convention on Access to Information, Public Participation in Decision-Making and Access to Justice in Environmental Matters, Aarhus, Denmark, June 25, 1998, http://www.unece .org/env/pp/documents/cep43e.pdf. See also D. B. Hunter, "The Emergence of Citizen Enforcement in International Organizations," in International Network for Environmental Compliance and Enforcement, Seventh International Conference on Environmental Compliance and Enforcement, April 9–15, 2005, *Conference Proceedings*, vol. 1, http://www .inece.org/conference/7/vol1/Hunter.pdf; and G.-S. Kremlis, "The Aarhus Convention and Its Implementation in the

European Community," in International Network for Environmental Compliance and Enforcement, Seventh International Conference on Environmental Compliance and Enforcement, April 9–15, 2005, *Conference Proceedings*, vol. 1, http://www.inece.org/conference/7/vol1/22_Kremlis .pdf.

5. A. Arato, *Civil Society, Constitution, and Legitimacy* (Lanham, MD: Rowman and Littlefield, 2000). J. Cohen and A. Arato, *Civil Society and Political Theory* (Cambridge, MA: MIT Press, 1994).

6. R. Westbrook, *John Dewey and American Democracy* (Ithaca, NY: Cornell University Press, 1991), xv.

7. It must be said that while democracy respects and values all groups, not all groups value democracy. The ideals of equality, inclusiveness, and solidarity do not fit well with the traditional beliefs and practices of many religions and cultural groups, so initiatives to promote ecological citizenship, particularly those that are based on a deliberative procedure, may not be readily embraced. The reasons for this reluctance may be insightful and deep. They may go beyond the sheer complexity of the subject matter, and its seeming distance or irrelevance to the community. And they may go beyond historical mistrust and suspicion that some communities feel about something that is perceived to be brought in by outsiders. In addition, there may be a sense that the purpose of these meetings is not only to inform or empower the members of the community, but also to transform them morally and politically. This suspicion is not without foundation.

8. See D. Mayhew, *Divided We Govern*, 2nd ed. (New Haven, CT: Yale University Press, 2005). T. E. Mann and N. J. Ornstein, *The Broken Branch: How Congress Is Failing America and How to Get It Back on Track* (New York: Oxford University Press, 2006). T. Frank, *What's the Matter with Kansas?* (New York: Henry Holt, 2004).

9. D. Matthews, *The Ecology of Democracy* (Dayton, OH: Kettering Foundation Press, 2014).

Part IV: The Political Economy of Climate Change—Democracy, If We Can Keep It

1. For this quotation I have followed the George Lawrence translation. See A. de Tocqueville, *Democracy in America*, ed. J. P. Mayer (New York: Anchor Books, 1969), 703.
2. Pope Francis, *Laudato Si': On Care for Our Common Home*, §111.

11. The Ecological Contract and Climate Change

1. J. M. Broder, "What's Rotten for Obama in Denmark," *New York Times*, December 13, 2009, WK 1, 4.
2. See http://www.bartleby.com/73/1593.html.
3. C. Hamilton, *Earthmasters: The Dawn of the Age of Climate Engineering* (New Haven, CT: Yale University Press, 2013).
4. C. W. Mills, *The Sociological Imagination* (New York: Oxford University Press, 1959), 8.
5. W. Nordhaus, *The Climate Casino: Risk, Uncertainty, and Economics for a Warming World* (New Haven, CT: Yale University Press, 2013).
6. J. Hansen, P. Kharecha, M. Sato, V. Masson-Delmotte, F. Ackerman et al., "Assessing 'Dangerous Climate Change': Required Reduction of Carbon Emissions to Protect Young People, Future Generations and Nature," *PLoS ONE* 8, no. 12 (2013): e81648. doi:10.1371/journal.pone.0081648.
7. G. Goldman, D. Bailin, P. Rogerson et al., *Toward an Evidence-Based Fracking Debate: Science, Democracy, and Community Right to Know in Unconventional Oil and Gas Development* (Cambridge, MA: Union of Concerned Scientists, 2013), www.ucsusa.org/assets/documents/center-for-science-and-democracy/fracking-report-full.pdf.
8. Center for Health and the Global Environment, Harvard Medical School, *Climate Change Futures: Health, Ecological and Economic Dimensions* (Cambridge, MA: Center for Health and the Global Environment, 2005); WHO,

Millennium Ecosystem Assessment, Ecosystems and Human Well-Being: Health Synthesis (Geneva: WHO Press, 2005).

9. A. Haines and J. Patz, "Health Effects of Climate Change," *JAMA* 291, no. 1 (January 7, 2004): 99–103; H. Frumkin and A. J. McMichael, "Climate Change and Public Health: Thinking, Communicating, Acting," *American Journal of Preventive Medicine* 35, no. 5 (2008): 403–10.

10. Hansen et al., "Assessing 'Dangerous Climate Change,'" 8.

11. The Hartwell Paper, *A New Direction for Climate Policy after the Crash of 2009*, 2010, http://eprints.lse.ac.uk/27939/1/Hart wellPaper_English_version.pdf.

12. S. Weart, *The Discovery of Global Warming*, rev. and expanded ed. (Cambridge, MA: Harvard University Press, 2008).

13. J. Rockström, W. Steffen, K. Noone, A. Persson, F. S. Chapin III, E. F. Lambin, J. A. Foley, "Planetary Boundaries: Exploring the Safe Operating Space for Humanity," *Ecology and Society* 14, no. 2 (2009): 1–33.

14. N. Klein, "Capitalism versus the Climate," *The Nation*, November 9, 2011, http://www.thenation.com/article/164497 /capitalism-vs-climate; N. Klein, *This Changes Everything: Capitalism vs. the Climate* (New York: Simon and Schuster, 2014); A. Parr, *The Wrath of Capital: Neoliberalism and Climate Change Politics* (New York: Columbia University Press, 2013).

15. H. Jonas, *The Imperative of Responsibility: In Search of an Ethics for the Technological Age* (Chicago: University of Chicago Press, 1985).

16. E. Callenbach, *Ecotopia* (Berkeley, CA: Banyan Tree Books, 1975). W. Ophuls, *Plato's Revenge: Politics in the Age of Ecology* (Cambridge, MA: MIT Press, 2011).

17. Such an understanding is necessarily rooted in robust scientific investigation and inference, but science alone cannot define this standard. It must also be based on the enduring experiences and traditions of humankind,

as we can know them from historical and anthropological study.

18. S. M. Gardiner, *A Perfect Moral Storm: The Ethical Tragedy of Climate Change* (New York: Oxford University Press, 2013); D. A. Brown, *Climate Change Ethics: Navigating the Perfect Moral Storm* (New York: Routledge, 2013); and D. Jamieson, *Reason in a Dark Time* (New York: Oxford University Press, 2014).

19. D. Runciman, "How Messy It All Is," *London Review of Books*, October 22, 2009, 3–6.

20. P. A. Victor, *Managing Without Growth: Slower by Design, Not Disaster* (Cheltenham, UK: Edward Elgar, 2008).

12. An Inquiry into the Democratic Prospect

1. See G. M. Turner, "A Comparison of the Limits to Growth with 30 Years of Reality," *Global Environmental Change* 18 (2008): 397–411. G. M. Turner, *Is Global Collapse Imminent? An Updated Comparison of the Limits to Growth with Historical Data*, MSSI Research Paper no. 4 (Melbourne, Australia: Melbourne Sustainable Society Institute, The University of Melbourne, August 2014).

2. Cf. Jimmy Carter, "Crisis of Confidence," transcript of televised speech from July 15, 1979, *American Experience*, http://www.pbs.org/wgbh/americanexperience/features/primary-resources/carter-crisis/.

3. D. Harvey, *A Brief History of Neoliberalism* (Oxford, UK: Oxford University Press, 2005).

4. G. Speth, *The Bridge at the Edge of the World: Capitalism, the Environment, and Crossing from Crisis to Sustainability* (New Haven, CT: Yale University Press, 2008).

5. R. Wilkinson and K. Pickett, *The Spirit Level: Why Greater Equality Makes Us Stronger* (New York: Bloomsbury Press, 2009).

6. I am assuming here, perhaps prematurely, that history has passed its negative judgment on the experiment of state

socialism and complete public ownership and control of the entire economy.

7. P. Groenewegen, "English Marginalism: Jevons, Marshall, and Pigou," in W. J. Samuels, J. E. Biddle, and J. B. David, eds., *A Companion to the History of Economic Thought* (Malden, MA: Blackwell, 2007), 246–61.

8. M. O'Connor, ed., *Is Capitalism Sustainable? Political Economy and the Politics of Ecology* (New York: The Guilford Press, 2004).

9. W. Ophuls, *Ecology and the Politics of Scarcity* (San Francisco: Freeman, 1977); W. Ophuls, *Requiem for Modern Politics: The Tragedy of the Enlightenment and the Challenge of the New Millennium* (Boulder, CO: Westview Press, 1988); and W. Ophuls, *Plato's Revenge: Politics in the Age of Ecology* (Cambridge, MA: MIT Press, 2011). J. Dryzek, *Discursive Democracy: Politics, Policy and Political Science* (New York: Cambridge University Press, 1990), and J. Dryzek, *Deliberative Democracy and Beyond* (New York: Oxford University Press, 2000). K. Bosselmann, J. R. Engel, and Prue Taylor, *Governance for Sustainability: Issues, Challenges, Successes* (Gland, Switzerland: IUCN Publications Services, 2008). K. Bosselmann and J. R. Engel, eds., *The Earth Charter: A Framework for Global Governance* (Amsterdam: KIT Publishers, 2010). K. Bosselmann, *Earth Governance: Trusteeship of the Global Commons* (Cheltenham, UK: Edward Elgar Publishing, 2015). J. Purdy, *After Nature: A Politics for the Anthropocene* (Cambridge, MA: Harvard University Press, 2015).

10. It may be said that interest group democracy displays various types and variations along a continuum running from relatively more elite dominated to relatively more populist in its policy and legislative responsiveness to mass public opinion, which in many cases is perhaps more accurately seen as an aggregation of private opinions. On this, see generally Bachrach, *The Theory of Democratic Elitism*.

11. G. Callabresi and P. Bobbitt, *Tragic Choices* (New York: W. W. Norton, 1978).

12. J. Lie, "Global Climate Change and the Politics of Disaster," *Sustainability Science* 2 (2007): 233–36. M. Beeson, "The

Coming of Environmental Authoritarianism," *Environmental Politics* 19, no. 2 (March 2010): 276–94.

13. R. L. Heilbroner, *An Inquiry into the Human Prospect* (New York: Norton, 1975).

14. Ophuls, *Ecology and the Politics of Scarcity*, 167–210.

15. Ibid., 163.

16. L. H. Gunderson and C. S. Holling, eds., *Panarchy: Understanding Transformations in Human and Natural Systems* (Washington, DC: Island Press, 2002).

17. It is more common to refer to "deliberative democracy" in this kind of discussion. I don't think that deliberative and discursive have quite the same meaning, although they are often used as synonyms. In choosing to develop discursive democracy here I follow the work of John Dryzek, who argues that discursive democracy denotes a political practice that is "pluralistic in embracing the necessity to communicate across difference without erasing difference, reflexive in its questioning orientation to established traditions . . . , transnational in its capacity to extend across state boundaries into settings where there is no constitutional framework, ecological in terms of openness to communication with non-human nature, and dynamic in its openness to ever-changing constraints upon and opportunities for democratization" (Dryzek, *Deliberative Democracy and Beyond*, 3).

18. For a broad contrast between representative and participatory or deliberative democracy see B. R. Barber, *Strong Democracy: Participatory Politics for a New Age* (Berkeley: University of California Press, 1994). For additional analysis of deliberative democracy see, in addition to the works of Dryzek, J. Fishkin, *Democracy and Deliberation: New Directions for Democratic Reforms* (New Haven, CT: Yale University Press, 1991); A. Gutmann and D. Thompson, *Democracy and Disagreement* (Cambridge, MA: Harvard University Press, 1996); W. F. Baber and R. V. Bartlett, *Deliberative Environmental Politics: Democracy and Ecological Rationality* (Cambridge, MA: MIT Press, 2005). For an overview of recent deliberative and participatory democratic theory, see F. Fischer, *Democracy*

and Expertise: Reorienting Policy Inquiry (Oxford, UK: Oxford University Press, 2009).

19. R. D. Wolff, *Democracy at Work: A Cure for Capitalism* (San Francisco: Haymarket Books, 2012), and R. D. Wolff and D. Barsamian, *Occupy the Economy: Challenging Capitalism* (San Francisco: City Lights Publishers, 2012).

20. N. Rogers, "Law and Liberty in a Time of Climate Change," *Public Space* 4 (2009): 1–34.

21. A. Demirović, "Ecological Crisis and the Future of Democracy," in M. O'Connor, ed., *Is Capitalism Sustainable? Political Economy and the Politics of Ecology* (New York: Guilford Press, 1994), 253–74; G. Garver, "The Rule of Ecological Law: A Transformative Legal and Institutional Framework for the Human-Earth Relationship" (master of law thesis, McGill University Faculty of Law, 2011).

22. J. A. Rohr, *To Run a Constitution: The Legitimacy of the Administrative State* (Lawrence: University Press of Kansas, 1986).

23. An analogue of ecological constitutionalism can be found in Jürgen Habermas's notion of "constitutional patriotism," a concept he developed in the context of post–Nazi Germany to sidestep the dangerous kind of nationalist loyalty that had proved so disastrous in the past, without abandoning the necessary motivational commitments that any form of governance needs because it cannot rely for solidarity or obedience on purely impersonal, transcendental moral arguments alone. In addition, unlike the nation-state, the notion of a constitution or a governance structure embodying certain substantive and procedural norms and rights can lend itself to both more micro-governance at local or ecosystemic levels and to macro-governance at the international level, such as the European Union. See J. Habermas, *The Post National Constellation* (Cambridge, MA: MIT Press, 2001), and J. Habermas, *The New Conservatism: Cultural Criticism and the Historians' Debate* (Cambridge, MA: MIT Press, 1992). See also J.-W. Müller, *Constitutional Patriotism* (Princeton, NJ: Princeton University Press, 2007).

24. J.-W. Müller, "Rule-Breaking: The Problems of the Eurozone," *London Review of Books*, August 27, 2015, 3–7.

25. I regard John Dewey as still one of the best examples of a democratic perspective that can be drawn on with much theoretical profit today. Cf. J. Dewey, *Individualism Old and New* (Amherst, NY: Prometheus Books, 1999), and J. Dewey, *Liberalism and Social Action* (Amherst, NY: Prometheus Books, 2000). For a commentary on Dewey's thought, see R. A. Westbrook, *John Dewey and American Democracy* (Ithaca, NY: Cornell University Press, 1993), and for democratic theorizing in this tradition, see R. A. Westbrook, *Democratic Hope: Pragmatism and the Politics of Truth* (Ithaca, NY: Cornell University Press, 2005), and J. Stout, *Democracy and Tradition* (Princeton, NJ: Princeton University Press, 2004). For a good historical overview of the changing shape of democratic theory within the individualistic liberal tradition, see C. B. Macpherson, *The Life and Times of Liberal Democracy* (Oxford, UK: Oxford University Press, 1977).

26. B. R. Barber, *Strong Democracy: Participatory Politics for a New Age* (Berkeley, CA: University of California Press, 1984).

27. S. MacGregor, "Reading the Earth Charter: Cosmopolitan Environmental Citizenship or Light Green Politics as Usual?" *Ethics, Place and Environment* 7, no. 1–2 (March–June 2004): 85–96.

28. R. A. Dahl and E. R. Tufte, *Size and Democracy* (Palo Alto, CA: Stanford University Press, 1973).

29. D. Villa, *Public Freedom* (Princeton, NJ: Princeton University Press, 2008).

30. Dryzek, *Deliberative Democracy and Beyond*.

31. See A. Guttmann and D. Thompson, *Democracy and Disagreement* (Cambridge, MA: Harvard University Press, 1996) for a discussion of criteria for assessing the adequacy of public debate in a democracy.

32. J. Cleese and G. Chapman, "The Argument Clinic" (1972), https://en.wikipedia.org/wiki/Argument_Clinic.

33. C. Hamilton, *Requiem for a Species: Why We Resist the Truth About Climate Change* (New York: Earthscan, 2010).

34. Leopold, *A Sand County Almanac*, 204.

Conclusion

1. B. McKibben, *Enough: Staying Human in an Engineered Age* (New York: Henry Holt, 2003).

2. E. Kolbert, *The Sixth Extinction: An Unnatural History* (New York: Henry Holt, 2014).

3. M. Weber, *The Protestant Ethic and the Spirit of Capitalism* (1905; New York: Charles Scribner's Sons, 1958). See also M. Horkheimer and T. W. Adorno, *Dialectic of Enlightenment*, trans. J. Cumming (New York: Herder and Herder, 1972).

Index

About the Author

Bruce Jennings is Senior Fellow at the Center for Humans and Nature, Adjunct Associate Professor at Vanderbilt University, and Senior Advisor at the Hastings Center, where he was a research scholar and served as executive director from 1990 to 1999. He has written widely on bioethics and public policy issues. He is editor-in-chief of *Bioethics,* 4th ed. (formerly the *Encyclopedia of Bioethics*), 6 vols. (Farmington Hills, MI: Macmillan Reference USA, 2014).

CPSIA information can be obtained
at www.ICGtesting.com
Printed in the USA
LVOW04s0747040516
486536LV00015B/75/P